BEATING MACULAR DEGENERATION WITH NUTRITION

Empower Yourself With Nutritional and Lifestyle Principles for Healing

Gluten-Free, Food Allergy Friendly, Diabetic and "Normal" Recipes

Nicolette M. Dumke

BEATING MACULAR DEGENERATION
WITH NUTRITION
EMPOWER YOURSELF WITH NUTRITIONAL
AND LIFESTYLE PRINCIPLES FOR HEALING

Published by
Allergy Adapt, Inc.
1877 Polk Avenue
Louisville, Colorado 80027
303-666-8253

Publisher's Cataloging-in-Publication data

Names: Dumke, Nicolette M., author.
Title: Beating macular degeneration with nutrition : empower yourself with nutritional and lifestyle principles for healing / Nicolette M. Dumke.
Description: "Gluten-free, food allergy friendly, diabetic, and 'normal' recipes" | Includes bibliographical references and index. | Louisville, CO: Allergy Adapt, Inc., 2019
Identifiers: LCCN: 2018957808 | ISBN: 978-1-887624-23-7
Subjects: LCSH Retinal degeneration--Alternative treatment. | Retinal degeneration--Diet therapy--Recipes. | Retinal degeneration--Prevention. | Cookbooks. | BISAC HEALTH & FITNESS / Vision | HEALTH & FITNESS / Diet & Nutrition / General | HEALTH & FITNESS / Allergies | HEALTH & FITNESS / Alternative Therapies | COOKING / Health & Healing / General
Classification: LCC RE661.D3 .D86 2018 | DDC 641.5/63--dc23

Dedication

To my dear Aunt Landa
(Yolanda Caldwell)
who is legally blind due to
macular degeneration
yet her sweet, loving and
always-positive spirit remains,

to Mark, my husband and hero,
for making innumerable changes
and cheerfully following all
the principles in this book
as his vision improves,

and, with endless praise, to God
who holds them both in His hands.

Disclaimer

The information contained in this book is merely intended to communicate food preparation material and information about possible treatment options which are helpful and educational to the reader. It is not intended to replace medical diagnosis or treatment, but rather to provide information and recipes which may be helpful in implementing a diet prescribed by your doctor. Please consult your physician for medical advice before embarking on any treatment or changing your diet.

The author and publisher declare that to the best of their knowledge all material in this book is accurate; however, although unknown to the author and publisher, some recipes may contain ingredients which may be harmful to some people and some treatments and natural remedies may be harmful to some people.

There are no warranties which extend beyond the educational nature of this book, either expressed or implied, including, but not limited to, the implied warranties of merchantability, fitness for a particular purpose, or non-infringement. Therefore, the author and publisher shall have neither liability nor responsibility to any person with respect to any loss or damage alleged to be caused, directly or indirectly, by the information contained in this book.

If you do not wish to be bound by the above, you may return this book to the publisher for a full refund.

Website

www.healingbasics.life/AMD.html

Contact

http://healingbasics.life/contact.html

Spread the Word!

Almost two decades ago, I heard a message that lay dormant in my mind until my family desperately needed it last year. At the time, I was listening to a conversation between my mother and her allergy doctor, and my mother mentioned that she had just been diagnosed with early-stage dry macular degeneration. The doctor's response was something very few people with macular degeneration ever hear. He told my mother that there was a nutritional intravenous therapy protocol for macular degeneration that "stops it in its tracks." Treatment with a highly nutritious diet and supplements can also stop the progression of macular degeneration, especially if started soon after it is diagnosed.

Without this information, my husband, Mark, would probably be experiencing progressively worsening vision now, possibly suffering the same extremely rapid progression of the disease his father had. Instead Mark now says things such as, "Back when *I thought I was going to go blind,* I wanted to see the world's best art while I could still see." At the first band concert in the parks this summer, he said, "There sure are a lot of blue chairs and hats here that I didn't see last summer." Instead of rapidly worsening, his vision has improved, all because of that one sentence I heard about eighteen years ago.

In their 1998 book, *Save Your Sight!* ophthalmologists Marc and Michael Rose, MD, who treat macular degeneration with food and supplements rather than IVs, write, "We have seen hundreds of patients halt the progression of macular degeneration and dozens actually reverse it. Most doctors don't yet embrace the kind of nutritional therapies in this book."[1] Twenty years after their book was published, nutritional therapy for macular degeneration is still almost totally unknown.

I often wonder – What if I hadn't heard that one sentence years ago? What about other spouses and families of macular degeneration patients and especially the patients themselves? They probably have not heard that sentence, and lack of that knowledge may make severe vision loss inevitable for them. I cannot remain silent while other patients and their families suffer. I want every macular degeneration patient and family to know that there is something that can help.

Readers of this book, **please help spread the word!** Tell all of your friends, relatives, co-workers and everyone you know who is 60 or older that there is hope. **Take this book to your library** and show it to the librarians. Tell them to use the contact form at www.healingbasics.life/contact to request a complimentary e-book of *Beating Macular Degeneration with Nutrition*. **Show this book to your audiologist, senior center staff** and **anyone else who serves seniors**. Tell them to use the contact form at www.healingbasics.life/contact to request a complimentary office copy of this book for their waiting rooms.

Spread the word! The person whose vision you save will be very grateful that you did not keep this information to yourself.

Thank you for helping prevent the tragedy of blindness in the lives of people who need to hear what I heard long ago.

This section is set in a different type style in anticipation of a large print edition of this book. We would appreciate hearing what you or your loved one with macular degeneration think about ease of reading the type above. Contact information is at the bottom of page 4.

---

1 Rose, Marc R., MD and Michael R. Rose, MD. *Save Your Sight! Natural Ways to Prevent and Reverse Macular Degeneration.* (New York, NY: Warner Books, 1998), xii.

Table of Contents

Hope

I once heard a speaker say that we can live a few weeks without food, a few days without water, a few minutes without air, but only a few seconds without hope. Hope is indeed essential to life, to living rather than just surviving or existing.

This is a story of hope. However, it did not begin that way. The message we originally received was, "Macular degeneration? We have new drugs for wet macular degeneration." This wasn't reassuring because we are very close to my aunt who has been treated with these drugs for over thirteen years and became legally blind during that time. No hope for her, just progressive loss of sight.

The week after my husband, Mark, was diagnosed with macular degeneration at a routine refraction appointment, he had an appointment with a retina specialist. She told us that his dry macular degeneration was already in the intermediate stage and gave us no hope except for telling us that when his condition progressed to the wet form of macular degeneration, there were drugs that could be injected into the eye that could slow the progress of the disease. This treatment option was devoid of hope for us because my aunt has taken (and still takes) over a hundred injections in her right (wet) eye and yet the disease progressed to a degree of blindness that makes it impossible for her to live independently at home. She is also legally blind in her left eye which is affected with advanced dry macular degeneration.

Our hope came from something I had heard eighteen years previously when my mother was diagnosed with early stage dry macular disease. I was listening when she was told by her allergy doctor, "There is an intravenous (IV) therapy protocol for macular degeneration that stops it in its tracks."[1]

One week after we saw the retina specialist, Mark had an appointment with a naturopath at a clinic that could give him the macular degeneration IV protocol. Rather than delaying giving Mark help while the clinic obtained the necessary solutions for the IVs, the doctor got him started immediately on eating an intensely nutritious diet. She even knew how to get foods he did not like into him. Her "Eye Smoothie" (recipe on page 87) contains kale and fish oil, and he doesn't mind them that way. He also began taking more supplements, getting more exercise, and drinking more water. When his blood tests results came back, his diet became even better for his eyes. His food choices were changed due to his hemoglobin A1c blood test result being in the pre-diabetes range. Our new goal became to protect his eyes from blood sugar surges

---

1 Not every macular degeneration patient needs to take IVs to arrest the disease. Early stage dry macular degeneration, especially, responds well to an intensively nutritious diet and supplements. In *Save Your Sight!* ophthalmologists Marc and Michael Rose, MD, who treat macular degeneration with food and supplements rather than IVs, write, "We have seen hundreds of patients halt the progression of macular degeneration and dozens actually reverse it." Quote from: Rose, Marc R., MD and Michael R. Rose, MD. *Save Your Sight! Natural Ways to Prevent and Reverse Macular Degeneration.* (New York, NY: Warner Books, 1998), xii.

When the IVs began, he noticed a small improvement in some aspect of his vision every week. At first he noticed things at night such as a line of car headlights extending for miles while driving in the dark and the lighted railings at the water plant behind our house in the evening, which he thought were beautiful. The next week, he noticed that a family across the street had Venetian blinds on their patio door. He had never before detected the slats when light shone though the blinds in the evening. The following week headlights became *two* separate twinkling lights rather than one shining blur of light.

Then colors which he hadn't realized he wasn't seeing became intense and very enjoyable. He suddenly loved the color of a bright blue blouse that I'd had and worn in warm weather for two years. He hadn't realized that I owned a blouse that color before. Yellow traffic lights and school busses were bright orangish-yellow just as he remembered from elementary school years. The purple gem iris in our yard with white and bright orange or yellow markings in the centers were a delight to him. They had had bloomed unnoticed the previous spring.

The question that arose was, "Why is intensive nutritional treatment for macular degeneration almost unknown and rarely practiced?" I'm not sure there is a good answer to that question. Both eye doctors we saw initially recommended the AREDS™ formula supplement, which is woefully incomplete. (See more about this on pages 25 to 26). After they've told a patient to eat kale, salmon and egg yolks and take this supplement, their work on nutrition seems finished.

There is very little *good* nutritional advice for macular degeneration or diabetes to be found either in books or on the internet. It's all about eye injections, diabetes drugs, and American Diabetic Association diets that are so liberal and dependent on drugs that they almost insure that all diabetics suffer complications that lead to amputations, blindness, and heart disease.[2]

But take heart: there is hope! Macular degeneration does not have to be a sentence of progressive blindness. I fervently hope that this book will begin to remedy the lack of information on *real* (not AREDS™) nutritional treatment for macular degeneration.

Spread the word! Share the hope with everyone you know so they will have heard, "There is hope for macular degeneration" just as I had heard eighteen years before Mark's diagnosis. Then they will have that hope to help them explore the options in this book if they or someone they love is diagnosed with macular degeneration.

---

2 "The target blood sugar ranges for people with diabetes listed by both the Joslin Diabetes Center and the American Diabetes Association go as high as 130 mg/dl for fasting blood sugar and up to 180 mg/dl after meals or snacks. It is these levels that lead to widespread damage in people with 'controlled' diabetes… Debilitating, life-altering repercussions are expected in people with diabetes who are being treated according to current standards." Quote from Carpender, Dana. *The Low-Carb Diabetes Solution Cookbook.* (Beverly, MA: Fair Winds Press, 2016), 14.

An Unexpected Diagnosis

About a year ago, Mark needed a new prescription to get safety glasses for work. I offered to accompany him just in case he decided to get his eyes dilated and would need someone to drive him home. True confessions: Both of us have had undilated eye examinations routinely so we could drive immediately after the appointment and get back to doing what we needed to do quickly. Mark's reply to my offer was that he didn't plan to get his eyes dilated but wouldn't mind having me along for company.

He had the refraction and got his new prescription. Then the doctor convinced him to let her dilate his eyes. Normally, when she looks inside the eyes with a hand-held ophthalmoscope, she looks and looks and looks at one eye, and then moves to the other eye and does the same. At this appointment, I was sitting behind her, facing Mark, and saw that she looked at one eye for just a few seconds and then quickly moved to the second eye and examined it for her normal amount of time. Then she returned to the eye she had first looked at and examined it thoroughly.

She said, "I'm glad you came in today. You have early stage dry macular degeneration."

We were both stunned. Mark's father was diagnosed with macular degeneration when he was ten years older than Mark was that day, and within a few months his dad could barely see. My beloved 94-year old aunt had moved to an assisted living facility eleven months previously to continue recovering from a fall. She had hoped to return home soon. However, she is legally blind in spite of having taken over 100 injections of drugs in her wet eye. She keeps taking the injections, although less often than initially, because she thinks her vision will be worse if she doesn't.

I was visiting her one day when the physical therapist came. We all went for a walk in the hall of her building. During one of our rests, she was talking, as she often did, about how much she wanted to go home to her dog. The therapist asked her, "What do you see when you look at my face?" She replied, "I can see the edges of your head, but not your features. He said, "It's not safe for you to go home because you might not see something on the floor and could trip and have another fall."

Because of our family members' experiences, when we heard the eye diagnosis both Mark and I knew, immediately and with certainty, how macular degeneration can devastate a life.

The doctor advised Mark to eat kale every day, salmon once a week, four egg yolks a week for lutein and take the AREDS™ eye supplement. (See pages 25 to 26 for more about the AREDS™ supplements). She also said he should see a retina specialist and gave us the names of two doctors nearby. I asked her if either of them was holistic. She said, "With dry macular degeneration, nutrition is the only thing you can do. They both are holistic because both will give you nutritional advice."

When we arrived at home, I made an appointment with the doctor she had recommended most highly and who had all five-star ratings on internet reviews. Her patients praised her in the reviews for getting them in for treatment quickly in a crisis and said that her staff was very kind.

During the seemingly-long nine days between the routine eye examination and his appointment with the retina specialist, Mark and I discussed everything we had ever heard or had just read on the internet about macular degeneration. The most important thing we talked about was what I had heard about eighteen years previously. My 78-year-old mother had been diagnosed with early stage dry macular degeneration. She was taking allergy shots called EPD[1] which our son Joel and I also took. (We now take LDA, an Americanized version of the shots. For more information, see the footnote below). When she had a routine between-shots phone appointment with her allergy doctor, with me on another phone extension, she mentioned that she had just learned that she had macular degeneration. The doctor told her, "There is an IV protocol that stops it in its tracks. If it gets worse, you should take the IVs."

Mark's appointment with the retina specialist lasted more than two hours. First an assistant took a detailed health history, had Mark read the Snellen eye chart, and put dilating drops in his eyes. We sat in a mid-office waiting area while the drops took effect. Two types of scans followed. Finally we saw the doctor, and she gave us the results: Intermediate stage dry macular degeneration. Intermediate already? She gave us a handout with a list of foods he should eat including leafy green vegetables, with kale at the top as the best, salmon or other fish, and egg yolks and with information about the AREDS™ supplement. We asked her how fast Mark's macular degeneration might progress.

"Impossible to tell."

Did she have any other nutritional advice? We mentioned some information we had read.

"We have no evidence that any of those things help."

Mark said that he had been following a low glycemic index diet for several years. She said she did not think it would be helpful.

The one thing that she did know was that he should look at an Amsler grid daily with each eye, and if the lines looked wavy or a dark area appeared in the middle of the grid, he should call her as soon as possible during a weekday "but not at 4 pm on Friday afternoon." She would give him an injection of a drug in his eye within a few weekdays of his call. She gave us no real hope: she offered only the treatment that had failed to help my aunt.

It was time for Plan B. I Googled "IV therapy near me," and discovered a clinic 4.5 miles away. I called and made an appointment for an initial office visit. The receptionist told me that one of their doctors had experience in treating macular degeneration with IVs. I contacted the doctor who had told my mother about the IV protocol years before. Although retired, he emailed me the protocol within a few

---

1 EPD stands for Enzyme Potentiated Desensitization. The Americanized version of EPD is LDA, which stands for Low Dose Allergens. For more about these treatments see www.food-allergy.org/epd.html and http://www.drshrader.com/lda_therapy.htm .

hours. One week after the seeing the retina specialist, with the IV protocol in hand, Mark had his first appointment with a naturopath.

She told us that she had worked with an MD who used a similar protocol for macular degeneration and that it worked well for stopping the progression of the disease. She also told me how to get plenty of spinach and kale into Mark, plus fish oil that he previously could not stand to take, plus plenty of anthocyanins from blueberries plus other nutrients – all via a smoothie. She said that the Occudyne™ supplement that Mark had switched to when we read about the AREDS™ supplement online and became disillusioned with it was a good, complete eye supplement. She also said that the nutrients in Occudyne™ and in the IV protocol were in safe amounts. We discussed supplements Mark was already taking and should take and she advised us about them. We discussed whole grains, green leafy vegetables, red, orange, yellow and purple vegetables, tomatoes, blueberries, omega-3 fatty acids and vitamin D. She also told me which brand of fish oil to purchase for palatability and where to buy it. She said Mark should drink half his body weight in ounces of water per day and that he needed to relax and let his parasympathetic nervous system function as much of the time as possible. He said, "I need to become less compulsive about work."

She gave us lab paperwork for a vitamin D blood test and a battery of other relevant tests. We left the office feeing hopeful and happy to have heard that there really is something that can be done to limit or stop the progression of macular degeneration and even prevent blindness.

Most of the advice she gave us was about nutrition. Although Mark's nutrition had improved quite a bit during 41½ years of marriage, his childhood had not been nutritionally ideal, so there was catching up to do. I started a "cooking frenzy," as Mark called it, which was therapeutic for me. We both began to learn more about what could be done to help Mark's eyes. What we learned and experienced is in the following chapters.

Our Family Nutritional History

Eating habits developed in childhood usually follow us throughout life. Thus, each person's nutritional history has a lasting impact on health. This chapter is the history of Mark's and our family's nutritional journey. I am also including my nutritional "roots" because of their learning experiences and influence on Mark.

I consider myself blessed to have a good nutritional history. I came from a large extended family of cooks and gardeners who encouraged children to help in the kitchen from an early age. They also taught children to love plants by giving us our own little flower beds and mini-vegetable gardens. (This practice helped protect the "real garden." We were told to leave it alone and "work" in our mini-gardens).

Learning about edible plants also happened at least weekly when we visited our grandparents, aunts and uncles. Unless it was winter, the first thing that happened on a visit to any relative was that the men and children – and women also unless they were busy cooking – traipsed out to the garden to see what progress the vegetables and fruit trees had made since the last visit. We sampled vegetables and tree-ripened fruit after a quick rinse with the hose, and Grandpa made each child a zucchini whistle using a leaf stalk from a zucchini plant and his pocket knife. We knew and ate *real food* long before Michael Polan encouraged that practice.

I remember watching my grandfather graft two kinds of apple branches from his trees onto an apple tree in our yard, making pie crust and learning how to crimp the edge with my aunt, and Italian sweet bread baking lessons from my grandmother. I'm sorry to say that I didn't put most of the gardening lessons into practice in my own yard until just a few years ago. I often wish I had the older generations around to ask for garden advice on a regular basis, but now I'm the "old person" who passes the family cooking heritage and gardening memories down to my children. When our son John helps me garden, he often asks what my father did about weeds and other gardening challenges.

When I was young, children were involved in getting food from the garden to our plates regularly. I remember my dad appearing at the back door with grocery bags full of romaine, red or green leaf lettuce, spinach and other vegetables. I first began washing them when I still needed a chair to reach the kitchen sink. I learned at a young age to add salt to the water to kill bugs and then lift the greens, letting each handful drain, into fresh water on the other side of the sink. I remember summer evenings during my pre-teen through college years spent on the patio snapping grocery bags full of Italian beans with my dad while we talked. The next day we would all be involved in processing and freezing the beans.

When the fruit from our trees was harvested, my mother and I sat together at the kitchen sink peeling and talking for hours. We made applesauce and also canned peaches, pears and sweet dark purple Italian plums. Nobody looked for a no-work shortcut to healthy and delicious food. Getting food from the garden to the table involved the entire family, and memories of long conversations with my parents during food-related tasks are precious.

During the 1950s, girls especially were encouraged to spend time in the kitchen and learn cooking techniques and tricks from relatives and family friends. I began living in the kitchen at a young age. In my preschool years, I "helped" my mother bake. When she put the muffins or bread in the oven, she turned on the oven light and I sat in front of the oven watching dough rise and brown as it baked. Watching the transformation in the oven was more exciting than watching TV.

Our nutrition was controlled and was very important to my mother. We could snack on fruit, carrots, celery or fennel from the produce drawers of the refrigerator freely. When I was older, I would drizzle a stalk of celery or piece of fennel with olive oil and sprinkle it with salt and pepper, or fill a bowl with lettuce and small chunks of cheese for a snack. Dessert other than fruit, on the other hand, was reserved for special occasions, and candy was strictly rationed. The candy we received at Halloween and for Easter was kept in a high cupboard. On Saturday night, we could choose one piece, and after eating it, we brushed our teeth.

Mark's childhood food experience, unfortunately, was not as pleasant as mine. He enjoyed dinner when they had steak or a roast because those were plain foods. He did not like the open-and-pour casseroles[3], canned spaghetti, or pizza made from a box that they often had. He tried to hide how little he was actually eating of these foods and then freely snacked from the always-well-stocked candy drawer in the kitchen after and between meals. As he grew older, he learned to make himself lettuce and mayonnaise sandwiches and Cup-of-Soup™ meals.

When Mark met me he discovered an entirely new approach to food. He loved pizza made completely from scratch, and we had dates on which we made pizza together in my parents' kitchen. Large extended family dinners were a delight to him, not only because of the abundance of homemade and homegrown foods, but also because my relatives welcomed him into the family enthusiastically. Grandma called him "Marco" and always commented on his "blonde" hair. (Blonde by Italian standards only, it was actually light brown).

After we were married, he began teaching me about food. On our honeymoon, he introduced me to the pleasure of picking something from the restaurant's dessert cart every evening. The trend continued after we settled in the city where he was attending graduate school about 1000 miles away from our parents' homes. When my parents came to visit after we had been married about six months, my mother and I were working together in the kitchen. I dropped something on the floor and bent over to pick it up. My mother looked at my behind and said, "You better watch out. You're going to end up looking like..." and she mentioned the names of relatives who inherited the Savioli body type. (My paternal grandmother, Maria Savioli Jiannetti, was quite wide in spite of working on the family farm all day almost every day from spring through fall. Her brothers were wide and tall).

---

3 Open-and-pour casseroles were made from cans of cream of mushroom soup, canned meat or fish, and other unthinkable things, opened and poured into a casserole dish, mixed, sometimes sprinkled with crushed potato chips and then baked until warm.

I took my mother's warning seriously and tried to lose weight with a standard low calorie, low fat weight loss diet. I carried a small spiral notebook with me, recorded everything I ate plus the calorie count, and kept my food intake at 500 calories less than the books said I needed per day. According to the experts, I should have lost a pound a week, but I didn't, so I boosted my calorie deficit to 1000 calories per day. I began doing a lot of swimming and still didn't lose very much weight. In addition, I was starved all of the time.

An office assistant at work who was about 40 (which seemed old to me at the time) but was very slim and stylish routinely followed a high-protein, ultra-low-carbohydrate diet. I decided to try that. The first day I didn't feel that great by mid-afternoon, but I stuck with it. I lost weight but never really felt right. Then my uncle died and I flew home for the funeral and the weekend. For four days I ate normally, including bread, fruit, and foods I hadn't eaten at all for a few months. When I flew back, Mark picked me up at the airport. He put his arm around my waist and said, "Gained a little weight, didn't you?" It was that obvious! I had gained back every pound that I had lost.

Finally I found a book called *Low Blood Sugar and You* by Carlton Fredericks, PhD. I began to follow his diet which was balanced and contained a moderate amount of carbohydrate, yet not more than the equivalent of one slice of bread at any meal or snack. It directed that the dieter have a snack containing protein three times a day, mid-morning, mid-afternoon and at bedtime. I lost weight gradually and was never hungry. When I had slimmed down, I stopped strictly controlling portion sizes, but retained some of the basic habits from Dr. Fredericks' diet such as eating a protein snack when I was hungry between meals and listening to my body about what, when, and how much to eat. I also permanently gave up eating sugar and sweets and made fruit-sweetened desserts for special occasions such as birthdays and holidays.

When we were newlyweds, I also gave Mark food lessons, such as teaching him that vegetables other than canned peas and corn (the only vegetables his family routinely ate) were good if prepared correctly. He discovered that broccoli, which he remembered as strong-flavored, foul-smelling mush from a can, was delicious if cooked from the fresh vegetable. He learned to eat a wide variety of properly prepared vegetables such as beets, winter squash, cauliflower, broccoli, raw spinach, and asparagus.

I made some of the casserole-type foods that my mother had made such as stuffed peppers and cabbage rolls. Although I made these dishes from scratch using nutritious whole foods, Mark disliked casseroles of any kind. When we had cabbage rolls, he asked me to give him the cabbage, rice, meat and tomatoes separately the next time. I finally gave up on casseroles. When we visited our parents for holidays, my mother-in-law gave me well-intentioned advice on how to make life in the kitchen easier. I'll never forget one of her recipes: one can of Spam™ cut in cubes, one can of potatoes (I never knew that canned potatoes existed before that day!) also cut in cubes, a few tablespoons of pickle relish, one can of mushroom soup, or Miracle Whip™ mayonnaise could be used instead as another "glue" option. All the ingredients were mixed together in a casserole dish, sprinkled with crushed potato chips, and baked until warm. No wonder Mark didn't like casseroles!

Although I didn't eat sugar and developed food allergies after we'd been married a few years, I continued to cook what Mark liked. The main and side dishes were nutritious and made from scratch. I also made him desserts from scratch with sugar, although occasionally I'd use a fruit sweetener like apple juice concentrate or dates.

After our first child was born, Mark made a great realization. It was, "This kid needs me and will need me for a long time! I have to take care of myself." He wanted to lose weight. I persuaded him, from my experience with Carlton Frederick's weight loss plan, that giving up sugar might be a good start, which he did, but he made no other changes. He lost 35 pounds. I also tried to get him to eat protein-containing between meal snacks, but the squares of cheese I put in his work lunches for snacks were eaten at the same time as his sandwich or cold chicken. Eating breakfast on weekdays was something that took him several years to learn. I think our firstborn was in late elementary school when Mark finally began eating breakfast every day.

He eventually realized that his highs and lows of energy and mood were part of a low blood sugar problem and got serious about controlling it. When I read *The Insulin Resistance Diet* by Dr. Cherlye Hart, MD, and other books, I learned about the glycemic index and "upgraded" Dr. Fredericks' diet that I had followed thirty years previously to take advantage of the advances in medical science and be something that Mark could follow easily and be satisfied on. With Dr. Hart's program, he stayed within 30 grams of carbohydrates per meal balanced with protein most of the time, but occasionally one of us would "mess up."

I miscalculated how much spaghetti was two units of carbohydrate (30 grams) and discovered the error when I began re-checking the carbohydrate math for his meals and weighing the carbohydrate foods he ate after the macular degeneration diagnosis. Thankfully, we only had spaghetti once every three weeks.

He "messed up" occasionally when he did something like finding an open bag of potato chips in our son John's car while riding with him and eating enough chips to satisfy him fully. After that he felt destabilized for several days, which was something he forgot the next time he saw an open bag of chips. When we heard that his hemoglobin A1c blood test result was in the pre-diabetic range, we began striving for perfection in what he ate. I began weighing all of his carbohydrate portions rather than relying on previous weighings and estimating portion sizes from memory.

The naturopath had told Mark that he should eliminate all white flour, white rice and white potatoes. The rice was easy; he rarely ate rice. The white flour was more of a challenge because he hated 100% stoneground whole wheat bread when he started Dr. Hart's link-and-balance plan. We switched to sourdough bread made with white flour when I learned that sourdough has a glycemic index lower than stoneground whole wheat bread. The naturopath thought my two whole wheat sourdough recipes (on pages 139 and 141), which contain mostly whole wheat flour but also bread flour, were all right. Initially I added stevia to take the bitter edge off the whole wheat which Mark tasted, but now that he has become more accustomed to whole wheat, he likes whole wheat sourdough bread without stevia. In addition, I usually use King Arthur™ white whole wheat flour. It is milled from a different strain of whole wheat that is equal in nutrition to darker whole wheat flour (although its color is lighter) because the whole grain is included in grinding the grain into flour. King Arthur™ calls it "lighter in color and flavor." This white whole wheat flour also has a great advantage because it contains little or no glyphosate (RoundUp™)[4] although it costs the price of an "ordinary" bag of King Arthur™ flour rather than the higher organic flour price.

The "no white potatoes" was also a challenge. My first try on "alternative" potatoes was made with Jersey sweet potatoes. (They are called white sweet potatoes although their flesh is yellow). When boiled, Jerseys have a glycemic index of 44, well down in the "low" range, with less than half the GI of white bread. I boiled and mashed the sweet potatoes so that we could both eat them (with my dairy allergy) using only the potatoes, some of the cooking water, olive oil, and salt. This was way too much like "fake mashed potatoes" for Mark. In a few weeks, I again made mashed Jersey sweet potatoes but mashed them with milk and butter. He liked them that way with a freshly cooked turkey and the "turkey juice" poured over the turkey and potatoes. (Not only is turkey juice free of white flour but, unlike gravy, it requires no extra effort to make). I thought we had come up with a mashed potatoes solution for Thanksgiving. However, he hasn't consented to eat mashed white sweet potatoes again yet. See the recipes for these potatoes and mashed rutabagas, which he liked better, on page 115.

Another dietary challenge was the advice of the two eye doctors he saw initially. They said he needed to eat salmon, eggs, and kale (preferably) or other green leafy vegetables. The salmon and eggs were not much of a problem, but Mark really did not like kale, even when I cooked it with bacon. I bought some kale chips that he thought he could eat. Hoping that he'd snack on them in large quantities, I made kale chips patterned after the brand he liked (recipe on page 162). He liked them better because they were made with milder-tasting, young and thinner kale leaves, but he still was not eating many of them. We supplemented the kale chips with a large salad of spinach, tomatoes, and red peppers every evening. The naturopath solved the kale problem with her smoothie which is packed with everything he needs – kale, spinach, fish oil, blueberries and more. See the recipe on page 87.

The doctor said he wouldn't notice the kale in the smoothie, but I had my doubts. The first time I made it, I used spinach as the only vegetable and added some vanilla and stevia. The second day, I made most of the serving as I had the previous day and put it in a large glass. However, I reserved a small amount of the smoothie in

---

4 Quote from a customer care representative's email about King Arthur Flour™'s white whole wheat flour, their first identity preserved product: "In addition to other requirements of this [identity preserved] program, our farmers are not permitted to use glyphosate as a pre-harvest application on the white winter wheat it's milled from. Our team is continuing to expand the level of control we have over aspects of growing practices to our other products, and we recognize that our work in this is never done. Although most of our flour is from wheat fields and farmers that do not use glyphosate pre-harvest, we're unable to offer a 100% guarantee of this except on our organic line of flours."

the blender and added some kale to it, blended it and put in a small glass. He drank them both and asked, "What's in this little glass? It's better than the large glass." I was amazed! The doctor was right! Also, our bodies are right when we listen to them. Our taste buds are really sensitive to the nutrients we need the most.

Among the other changes Mark made was to drink half his body weight in ounces of water each day. We had purchased a water filter about 1¾ years previously when I learned post-mastectomy that water purchased in plastic bottles contains chemicals that leach from the plastic which could cause cancer to return. At that time, Mark insisted that he didn't need special water. When he began drinking more water, he switched to filtered water.

He also began eating a much wider variety of vegetables with the emphasis on highly colored types – bright orange kabocha squash, purple cabbage, beets, and his salads of spinach, tomatoes and red bell peppers. Peas, his old staple, appear on our dinner table only occasionally now.

A little over two months into the new way of eating, Mark decided that eight ounces of cottage cheese, a thin slice of whole wheat toast with butter, and a smoothie was too much food for breakfast and eliminated the toast. He did well and was less hungry later in the day. (This was probably due reducing breakfast carbo-hydrate from two to one carbohydrate unit which comes from the blueberries in the smoothie). After about three weeks, however, he was craving bread, so now he has a little sourdough whole wheat bread with high-protein dinner entrées occasionally.

After three months of no white flour, rice or potatoes, the super-nutrient dense diet, and strictly following Dr. Hart's eating plan presented in *The Insulin Resistance Diet,* he had another hemoglobin A1c blood test. It had gone down only slightly, so we began using a glucometer to monitor his blood glucose levels at two hours after meals. He passed the first test, a meal containing a hamburger with a sourdough bun, with flying colors and a two-hour post-meal blood sugar test result of 91 mg/dl. The next week he ate a moderately sized portion of high protein lasagne (recipe on page 110) with a two-hour postprandial glucose result of 98 mg/dl. A week later, he ate a generous portion of sourdough pizza and his two hours-after-dinner blood glucose test result was 98 mg/dl. I wondered if the glucometer was stuck on 98, so tested myself, and the result was different.

The next test, done about two weeks later, was an intentional challenge of a meal containing as much spaghetti as he desired. It contained about 60 grams of carbohydrate, twice as much as he'd eaten in a long time, except on rare occasions when he was a dinner guest. The result of his blood test two hours after this feast was 100 mg/dl. His glucometer readings were nowhere near the upper limit of 140 mg/dl for normal two-hour postprandial blood sugar test results. How could he be pre-diabetic and have such a low test result after eating that much carbohydrate in one sitting?

I searched the internet and learned that the hemoglobin A1c blood test is not the most reliable way to tell if and how much of a problem one has, but that two-hour postprandial glucose test results are more meaningful.[5] The hemoglobin A1c blood test assumes that the lifespan of the red blood cells is three months. However, for non-diabetics, the life span of red blood cells can be up to 150 days, or about five months. This allows the red blood cells about 66 percent more time to absorb glucose, which skews the test results. The faulty assumption built into normal hemoglobin A1c values is that everyone's red blood cells have a three-month lifespan. Also, taking 1000 milligrams of vitamin C per day, which Mark has done for years, can skew hemoglobin A1c test results.[6]

Although we were relieved that Mark's glucometer results showed that he does not have blood sugar surges (but rather led to the discovery that hemoglobin A1c test results may be meaningless), he found the "deprivation" that he had endured due to misleading blood tests discouraging. Dread of his upcoming six-month appointment with the retina specialist also took a toll on him mentally. He so strongly expected to receive bad news that, before hearing any news, he felt like he had given up pleasures of eating for nothing.

I tried to add some of the "fun" back to his diet by developing a cola recipe without sugar or artificial sweeteners (recipe on page 91) to replace the diet sodas he had given up. I also began making stevia or monk fruit sweetened cacao candy and cacao chip cookies (recipes on pages 172 and 153). This seemed to cheer him up. He felt better physically and thought it was because the cacao provided needed nutrients. I read that cacao is high in magnesium and antioxidants and also stimulates the production of neurotransmitters which can improve mood. [7]

When the appointment with the retina specialist finally occurred, it was a surprisingly positive experience. (See pages 71 to 72 for the details). Mark is now cheerfully eating everything he should. He enjoys his food as well as improved vision, the happy result of receiving superior eye nutrition.

---

5 Kresser, Chris. "Why Hemoglobin A1c Is Not a Reliable Marker." March 1, 2011. https://chriskresser.com/why-hemoglobin-a1c-is-not-a-reliable-marker/

6 Radin, Michael S. MD. "Pitfalls in Hemoglobin A1c Measurement: When Results may be Misleading." *Journal of General Internal Medicine.* 2014 Feb; 29(2): 388-394. https://www.springermedizin.de/pitfalls-in-hemoglobin-a1c-measurement-when-results-may-be-misle/9021322

7 Mercola, Joseph, DO. Dark Chocolate Reduces Stress and Inflammation, Boosts Memory and Mood. May 10, 2018. https://articles.mercola.com/sites/articles/archive/2018/05/10/dark-chocolate-benefits.aspx

About Macular Degeneration

Macular degeneration is a progressive disease in which the center part of the retina, called the macula, gradually dies, initially a few cells at a time. In a chapter title of his book, one ophthalmologist author called macular degeneration "Starvation of the Retina." Nutrition is the key answer to macular degeneration because the disease really is starvation of this small part of the eye in a body that may seem well nourished but actually is not. This disease is often called age-related macular degeneration (AMD or ARMD) because its incidence increases with age.[1]

All macular degeneration starts as dry macular degeneration and may become wet (neovascular) macular degeneration if abnormal blood vessels which leak and/ or bleed, grow under, in, or over the macula. Wet macular degeneration may cause sudden severe changes in vision. Dry macular degeneration usually progresses more slowly but can be just as destructive if geographic atrophy (GA) develops. With geographic atrophy, regions of the retina waste away (atrophy) and die.[2]

Macular degeneration affects central vision but not peripheral vision, so it does not cause "total blindness," which means that no stimulus from light reaches the brain. However, it often causes "legal blindness" which is vision (with corrective lenses) of 20/200 or worse.[3] Macular degeneration can make reading, driving, watching TV, recognizing faces, many kinds of work, and living independently difficult or impossible.[4] AMD may cause the pleasures one intended to enjoy in later years to become only a dream.

A number of risk factors exist for developing macular degeneration. Age of over 60 years is a major risk factor, and the risk rises with increasing age. Family history of macular degeneration also increases risk, as do light colored eyes and Caucasian race. Other factors that are possibly more controllable are exposure to ultraviolet (UV) or high-energy blue light, cardiovascular disease, obesity, smoking, exposure to toxins and environmental pollution, and low blood levels of minerals and antioxidant vitamins such as vitamins A, D and E.[5]

The macula is the center part of the retina and is packed with light sensing structures called rods and cones. (For more detail about how rods and cones turn light into vision, see pages 23 to 24. Beneath the rods and cones is the retinal pigment epithelium (RPE), a membrane containing protective pigments derived from the diet. Beneath the RPE is the choroid, a layer that contains blood vessels. With age,

---

1 Buettner, Helmut, MD, Editor in Chief. *Mayo Clinic on Vision and Eye Health*. (Rochester, MN: Mayo Clinic, 2002), 137.

2 https://www.brightfocus.org/macular/article/what-geographic-atrophy

3 Buettner, 139.

4 Buettner, 137.

5 Rose, Marc R., MD and Michael R. Rose, MD. Save *Your Sight! Natural Ways to Prevent and Reverse Macular Degeneration*. (New York, NY: Warner Books, 1998), 18; Heier, Jeffrey S., MD. *100 Questions and Answers About Macular Degeneration*. (Boston, MA: Jones and Bartlett Publishers, 2011), 19; and Buettner, 142-143.

the RPE deteriorates and thins, thus lessening both the transport of nutrients to the macula and waste products out of the macula. The drusen seen in the eyes with macular degeneration are deposits of waste products. Dry macular degeneration is diagnosed based on the appearance of drusen. Furthermore, the thinning and deterioration of the RPE gives the retina a mottled look rather than a uniformly red appearance. In the final stages of advanced dry macular degeneration, the RPE may deteriorate to the point of no longer supplying the macula with nutrients or supporting its function in other ways, thus resulting in complete loss of central vision.[6]

Wet macular degeneration occurs in only 10 to 15% of cases but causes 90% of the incidents of severe vision loss. Macular degeneration becomes "wet" when new blood vessels grow from the choroid. These vessels leak fluid or blood and can cause sudden legal blindness. Patients are instructed to look at an Amsler grid daily to see if the lines become wavy or a dark spot develops, which means they are experiencing bleeding or leaking of blood vessels under the macula. Injections of anti-VEGF (anti-vascular endothelial growth factor) drugs into the vitreous humor of the eye may stop current bleeding but do not prevent the growth of new abnormal blood vessels.[7] Also, although the drug-treated blood vessels no longer leak or bleed, they are not functional. They become scar tissue which replaces normal tissue.[8]

What is the underlying cause of the eye's problem with disposing of waste products that initiates macular degeneration?[9] Lack of antioxidants in the diet[10] leads to oxidation in the eye that causes blockage of small blood vessels, resulting in decreased nutrient transport and waste removal.[11] When the lack of blood flow becomes severe enough, the body attempts to correct the problem by growing new blood vessels. However, the new vessels are very fragile and leak, bleed, or grow out of the choroid and over the macula.[12]

As with many chronic diseases, the most effective treatment is prevention or nutritional and other natural intervention early in the disease process. For advanced wet macular degeneration with actively bleeding or leaking blood vessels, conventional treatment with anti-VEGF drugs can help temporarily by stopping the bleeding. However, the drugs do not prevent new abnormal blood vessels from growing and leaking or bleeding nor do they prevent the formation of scar tissue. Thus, progressive destruction of the macula continues, although more slowly. In my opinion, individuals with macular degeneration can best improve their odds against blindness with excellent eye nutrition, ideally started before their macular degeneration has had much time to advance.

---

6 Buettner, 139.

7 Heier, 49, 64.

8 Buettner, 142.

9 Buettner, 141.

10 Rose and Rose, 56-57.

11 Buettner, 142 and Rose and Rose, 56-57.

12 Buettner, 142.

How Rods and Cones Produce Vision

In the discussion of rods and cones earlier in this chapter, more information about how our eyes are "fearfully and wonderfully made" was promised. Here is the amazing story of how our eyes process light to become complex images of the world around us.

We first heard about the incredible function of rods and cones one night when both of our sons were eating dinner with us, and we were discussing Mark's improvements in color vision. Our older son, Joel, an electrical engineer whose dissertation subject was image processing, told us that he had learned in graduate school about how the human eye processes images. He said that our retinas are populated with rods and some cones that serve special purposes including detecting motion, lines, edges, etc. We have three kinds of color-detecting cones: short, medium and long. The short cones detect blue light, the medium cones detect green, and the long cones detect red. Although the detection of yellow depends mostly on the short cones, the signal sent from the retina to the brain for yellow is a combination of information from all three types of cones. We see yellow when the brain compares the signals from short (yellow/blue) cones to signals from long (red) and medium (green) cones.

Each rod or cone has a nerve coming from it. These nerves work together to combine and interpret the signals from nearby rods and cones to form an image, such as "red stationary object." These "processed" signals then travel to the brain and are transformed into a visual image there. In addition to detecting light, the retina is actually also a neurological structure, like an extension of the optic nerve into the eye, when it combines signals from various rods and cones to produce part of an image.

We learned more about this from a paper Mark found online titled "Detection of Early Loss of Color Vision in Age-Related Macular Degeneration – With Emphasis on Drusen and Reticular Pseudodrusen."[13] In addition to covering some of the information we heard from Joel, the paper said that the short yellow-blue (YB) cones are the fewest in number, comprising about 8% of the cones in the retina and less in the macula, ranging from a high of 5% across the outer parts of the macula to 0.5% in the center of the macula. Therefore, when macular degeneration patients are tested for color vision, as their vision worsens, they experience changes in yellow-blue color sensitivity first because the short cones are fewest in number, so the loss of just a few of them is more noticeable.

In this study, loss of yellow-blue color sensitivity preceded and was slightly greater than the loss of red-green (RG) color sensitivity. The patients with the highest loss of both yellow-blue and red-green color sensitivity were the most likely to

13 Vermala, Roopa, Sobha Sivaprasad and John L Barbur. "Detection of Early Loss of Color Vision in Age-Related Macular Degenration – With Emphasis on Drusen and Reticular Pseudodrusen." *Investigative Ophthalmology & Visual Science.* August 2017, Vol. 58, BIO247-BIO254.

have their disease progress rapidly to wet macular degeneration or to geographic atrophy, the advanced stage of dry macular degeneration.

In Mark's experience, as he was progressively regaining color vision with each IV he took, he noticed blue first after about three IVs. (Read about the long-owned but never noticed bright blue blouse on page 10). His realization that yellow traffic lights, center-of-the-street lines and school busses were bright orange-ish yellow rather than pale lemon yellow came next after five IVs. Then he became able to see more subtle differences in color, such as that most car headlights emit the warm yellow-tinged light color of incandescent bulbs but than LED headlights look blueish white. He also began to recognize a very light beige or cream color in places which he thought were white before.

We think that his eyes may have contained some short cones that were struggling nutritionally and therefore not functioning well, but were still alive. As they received high levels of much-needed nutrients with each weekly IV, they became more functional. The body's ability to heal is amazing! His experience tells us that as long as there is life left in a cone or rod in the eye, there is hope for recovery of its function with intensive nutrition. Thus there is hope for improvement in vision.

Treating Macular Degeneration

A variety of treatments exist for macular degeneration. Retina specialists usually treat wet macular degeneration by injecting the eye with anti-VEGF (anti-vascular endothelial growth factor) drugs to stop leaking or bleeding from small blood vessels in or under the macula. Currently, the two most commonly used drugs are Lucentis™ (ranibizumab) and Avastin™ (bevacizumab). Avastin™ was developed to treat colon cancer and is used off-label for macular degeneration. It is much less expensive than other anti-VEGF drugs. The oldest drug, Macugen™ (Pegaptanib sodium), is rarely used because it is less effective than Lucentis™ and Avastin™.[1] Newer drugs include Eylea™ (aflibercept) and Opthea™ (opt-302).

Anti-VEGF drugs do stop bleeding but because they do not prevent the growth of new abnormal blood vessels, patients need repeated injections. Blood vessels that have been clotted by the drug remain and may form scar tissue. When enough scar tissue forms, it can create a dark spot in the center of visual field.[2] Other treatments for wet macular degeneration include photocoagulation and laser surgery.[3]

There are no drug treatments for dry macular degeneration. The advice conventional eye doctors give to patients with the dry form is to eat kale or other dark green leafy vegetables, salmon and eggs and take an AREDS™ (Age-Related Eye Disease Study) supplement for eye health. This supplement is favored because large clinical trials were performed with it. Conventional eye doctors seem to consider all other eye supplements inferior because their makers could not afford to sponsor large trials.

The AREDS and AREDS2 clinical trials showed that about 25% of subjects with intermediate stage dry macular degeneration had slower progression of their disease to the advanced stage. The AREDS™ and AREDS2™ supplements did not prevent macular degeneration, slow progression from the early to the intermediate stage, or slow progression of the advanced stage of the disease to legal blindness.[4] When the results from the first AREDS study were analyzed by genotype, two genotypes comprising 31% of the study participants had a greater chance of progression than those taking the placebo.[5] There was no benefit from taking AREDS™ in these patients, but instead the progression to advanced wet macular degeneration was accelerated as shown by the study footnoted at the bottom of this page.[6]

---

1 Anshel, Jeffrey, OD and Laura Stevens, M. Sci. *What You Must Know About Age-Related Macular Degeneration.* (Garden City Park, NY: Square One Publishers, 2018), 25.

2 Buettner, Helmut, MD, Editor in Chief. *Mayo Clinic on Vision and Eye Health.* (Rochester, MN: Mayo Clinic, 2002), 142

3 Buettner, 145-146 and Anshel and Stevens, 145-146.

4 Heier, Jeffrey S., MD. 100 Questions and Answers About Macular Degeneration. (Boston, MA: Jones and Bartlett Publishers, 2011), 40-41.

5 Anshel and Stevens, 59 and "AREDS Eye Supplements: Helpful or Harmful?" https://www.macularisk.com/amd-pharmacogenetics/areds-eye-supplements-helpful-or-harmful.html

6 Vavvas, Demetrios G., Kent W. Small, *et al.* "CFH and ARMS2 genetic risk determines progression to neovascular age-related macular degeneration after antioxidant and zinc supplementation." *Proceedings of the National Academy of Sciences, USA*, 2018 Jan 23;115(4):E696-E704. doi: 10.1073/pnas.1718059115. Epub 2018 Jan 8. https://www.ncbi.nlm.nih.gov/pubmed/29311295

In the AREDS2 study, the amount of zinc in the formula was reduced to 25 mg. The AREDS2 formula slowed the progression of intermediate stage macular degeneration to the advanced state in about 25% of patients (like the original AREDS™ supplement) but also lacked positive results for prevention of progression at other stages. In spite of the equal results with the lower amount of zinc, the amount of zinc in the formula was kept at 80 mg. Dr. Jeffery Anshel, OD, and Laura Stevens, MSci, co-authors of *What You Must Know About Age-Related Macular Degeneration,* feel that the lower dose of zinc is appropriate.[7]

The original AREDS™ supplement (Bausch & Lomb's Preservsion™ AREDS Eye Vitamin and Mineral Supplement which is still sold) contains 500 mg of vitamin C, 400 IU of vitamin E, 15 mg of beta-carotene, 80 mg of zinc, and 2 mg of copper. For the AREDS2 study, the beta-carotene was removed from the supplement and lutein and zeathanthin were added. The AREDS2™ supplement (Bausch & Lomb's Preservsion™ AREDS2 Eye Vitamin and Mineral Supplement) contains 500 mg of vitamin C, 400 IU of vitamin E, 80 mg of zinc, 2 mg of copper, 10 mg of lutein and 2 mg of zeaxanthin.[8]

In *The Right Dose: How to Take Vitamins and Minerals Safely*, Patricia Hausman, MS, discusses a study which measured the HDL "good" cholesterol levels of men who took 80 mg of zinc per day for five weeks. This higher-than-usual dose of zinc reduced their HDL levels by 25%. For individuals with macular degeneration, the goal is to protect small blood vessels in the eyes. Since HDL helps remove deposits from blood vessel walls, lowering the blood level of HDL would seem counterproductive for health of the blood vessels in the eyes.[9]

In *The Eye Care Revolution*, ophthalmologist Robert Abel, MD says that the high level of zinc in AREDS™ inhibits the absorption of calcium, magnesium, selenium, chromium, vanadium and other minerals needed by macular degeneration patients. He also states that the copper which is added to AREDS™ to balance the high level of zinc reduces the clearance of Alzheimer's disease-associated beta-amyloid in the brains of mice. The mice also exhibited a four-fold increase in abnormal capillaries in the brains. This indicates that excess copper may increase the risk of Alzheimer's disease

At Mark's appointment with the retina specialist six month's after his diagnosis, the doctor told us about and gave him a sample of a new AREDS2 supplement, Systane I-Caps™ Chewable Eye Vitamin and Mineral Supplement AREDS2, which is made by Alcon/Novartis Pharmaceuticals. This supplement contains the reduced amount of zinc used in the AREDS2 study, 25 mg. It also contains 500 mg of vitamin C, 400 IU of vitamin E, 2 mg of copper, 10 mg of lutein and 2 mg of zeaxanthin.

In addition to the problems with the AREDS™ supplements mentioned above, there are a large number of nutrients that promote eye health but are not included in the AREDS™ supplements.[10] See the next page for the list of nutrients which

---

7 Anshel and Stevens, 59.

8 Anshel and Stevens, 55, 57.

9 Hausman, Patricia MS. *The Right Dose: How to Take Vitamins and Minerals Safely.* (New York, NY: Ballantine Books, 1987), 390.

10 Abel, Robert, Jr. MD. *The Eye Care Revolution: Prevent and Reverse Common Vision Problems.* (New York, NY: Kensington Publishing Corp., 1999, 2014), 187 and Anshel and Stevens, 65.

ophthalmologists Marc and Michael Rose, MD recommend that their patients take. They say, "We're going to give you very detailed nutritional guidelines that we have seen prevent, halt, and sometimes even reverse the progress of macular degeneration... If you can commit to our Ten Steps to Restoring Vision and Vitality and the nutritional protocol for eye disease for at least nine months, you can almost certainly improve your vision. At the very least, you can keep it from getting worse."[11]

Nutritional treatments which are more complete than AREDS™ effectively address the causes of macular degeneration. Medical professionals such as naturopaths, who are specifically educated and trained to use nutritional methods, have much to offer for the treatment of macular degeneration – dietary advice, complete supplementation protocols and intravenous nutritional therapy. Since lack of nutrients reaching the retina is linked to the development of macular degeneration, comprehensive nutritional treatment actually addresses the underlying root problem rather than putting a temporary Band-Aid™ on symptoms as the anti-VEGF drug injections do. (To find a naturopath nearby, see "Sources," page 221).

Two of the pioneers of nutritional therapy for macular degeneration were Jonathan Wright, MD, and Alan Gaby, MD. They developed the Occydyne and Occudyne II™ eye supplements based on their experiences treating patients and they also gave more advanced macular degeneration patients nutritional IVs. Dr. Wright's Tahoma Clinic claims 70% effectiveness for stabilizing vision and reducing further loss in those with dry macular degeneration. In some cases even gains in vision are made.[12]

In *Save Your Sight!*, ophthalmologists Marc and Michael Rose recommend that macular degeneration patients take daily oral supplements containing:

Vitamin C – 2000 mg
Vitamin A – 10,000 IU
Lutein – 6 to 10 mg[13]
Zeaxanthin – 0.3 to 1 mg[12]
Magnesium – 300 to 500 mg at bedtime
Fish oil – 1 teaspoon, should contain rosemary, etc. to prevent rancidity
Vitamin E – 800 IU
Selenium – 200 mg
N-acetyl cysteine – 500 mg, on an empty stomach 2 to 3 times per day
Taurine[14] – 500 to 2000 mg between meals
Zinc – 15 to 30 mg
CoQ 10 – 30 to 200 mg

---

11 Rose, Marc R., MD and Michael R. Rose, MD. Save *Your Sight! Natural Ways to Prevent and Reverse Macular Degeneration.* (New York, NY: Warner Books, 1998), xiii.

12 http://tahomaclinic.com/2011/11/prevent-stop-or-reverse-vision-loss/ and http://tahomaclinic.com/treatment-for-dry-macular-degeneration/

13 Lutein and zeaxanthin are pigments that protect the macula.

14 Taurine is for blood vessel health. Take it separately from N-acetyl cysteine

Additionally, macular degeneration patients should have a blood test for vitamin D levels. If the result is low, supplement with enough vitamin D3 to bring the blood level to 50 to 60 ng/ml and maintain it there.

Two other nutrients that have benefited eye health in recent years since the publication of *Save Your Sight!* are bilberry and astaxanthin. Bilberry contains anthocyanins, improves capillary fragility, is anti-inflammatory, is a powerful antioxidant, and improves insulin resistance in diabetics. The recommended dosage is 50 to 125 mg taken twice a day.[15] For more about how the anthocyanins in bilberry and blueberries help insulin resistance, see the article cited below which is available online.[16]

Astaxanthin is the pigment that gives salmon its color. It is made by a species of algae called *Haematococcus pluvialis* which is consumed by krill, a microscopic crustacean, which is consumed by salmon. When choosing an astaxanthin supplement, look for one made from *H. pluvialis* because natural astaxanthin has significantly greater antioxidant capacity than if it is made in a laboratory from petrochemicals.[17]

Natural astaxanthin is the most powerful carotenoid antioxidant, is potently anti-inflammatory, crosses the blood-brain barrier and blood-retina barrier (making it especially helpful for macular degeneration), absorbs ultraviolet-B light (thus protecting the eyes from damage), and helps stabilize blood sugar, among other beneficial effects.[18] Recommended dosages range from 2 to 12 mg/day.

The Occudyne II™ eye supplement that Mark takes contains most of the nutrients listed above and also includes B vitamins and trace minerals including boron, chromium, manganese, selenium and vanadium. Mark's naturopath said Occudyne II™ was a good eye supplement and all the nutrients were at safe levels but that it was not a substitute for a highly nutritious diet, so it should be taken in addition to making dietary changes. "Food is best," she says. See pages 17 to 19 for her dietary recommendations for Mark. For recommendations from many sources, see pages 33 to 39.

---

15 Anshel and Stevens, 72-73 and Bornsek, SM, L ziburna, et al. "Bilberry and blueberry anthocyanins act as powerful intracellular antioxidants in mammalian cells."
Food Chemistry, 2012 Oct 15;134(4):1878-84. doi: 10.1016/j.foodchem.2012.03.092. Epub 2012 Mar 30. https://www.ncbi.nlm.nih.gov/pubmed/23442633 The article may be requested here: https://www.researchgate.net/publication/235739688_Bilberry_and_blueberry_anthocyanins_act_as_powerful_intracellular_antioxidants_in_mammalian_cells

16 Stull, April J. "Blueberries' Impact on Insulin Resistance and Glucose Intolerance." *Antioxidants.* 2016 Dec; 5(4): 44. http://www.mdpi.com/2076-3921/5/4/44 (full text article)

17 Shah, Mahfuzur, Yuanmel Liang, et al. "Astaxanthin-Producing Green Microalga*Haematococcus pluvialis*: From Single Cell to High Value Commercial Products." *Frontiers in Plant Science*, 2016; 7: 531.Published online 2016 Apr 28. https://www.ncbi.nlm.nih.gov/pmc/articles/PMC4848535/

18 Mercola, Joseph, DO. "Astaxanthin: The Most Powerful Nutrient Ever Discovered for Eye Health." November 23, 2010. https://articles.mercola.com/sites/articles/archive/2010/11/23/astaxanthin-the-eye-antioxidant-550-times-more-powerful-than-vitamin-e.aspx

Mark's macular degeneration was diagnosed at the intermediate stage and his father went from diagnosis to being legally blind in a few months. Thus, in addition to excellent nutrition from food and supplements, he chose to have more assertive treatment with nutritional IVs. All potential problems with absorption are bypassed by the IVs and higher levels of nutrients than can be achieved by oral supplementation bathe the eyes with what they need effectively. The IV protocol he took is the most comprehensive of any I've seen. (Admittedly, I've only see descriptions of a few on the internet). It contains vitamin C, a multi-B vitamin complex, additional B-6 and B-12, magnesium, selenium, zinc, boron, copper, molybdenum, chromium, manganese, vanadium, lithium, strontium, and taurine.

Mark and I first heard about astaxanthin after he had completed treatment with nutritional IVs. He was very impressed with what he read, and I told him I could easily add it to his smoothies if he didn't want yet another supplement to swallow. Then I added it without mentioning it. On the fourth day after he consumed the first astaxanthin-containing smoothie, he saw something he had never before seen in the Magic Eye™ 3D image books he received as gifts several months previously. He discovered that the designs on the inside of the front and back covers of the books were also 3D images. He hadn't seen anything totally unexpected since the third month he took IVs so was wondering what changed. When I told him that I began adding astaxanthin to his smoothies a few days earlier, he was sure that was what made him able to see 3D much more easily. A week later, he looked at a Magic Eye™ image that is mostly shades of blue. He had never been able to make out a hidden object in this image before, but after about ten days of taking astaxanthin, an old Spanish sailing ship popped out at him. For more about all the improvements in vision he experienced, see pages 10 and 70 to 72.

Nutritional treatment for macular degeneration may succeed after conventional treatment has failed. In *Blind Faith*,[19] John Crittenden tells of consecutively taking injections of three different anti-VEGF drugs monthly for a total of nineteen months with no improvement in his vision. In fact, during this time, he became legally blind. Then, with his ophthalmologist's blessing and with assistance from a naturopath, he researched and worked out a diet and supplement protocol that restored his vision from 20/400 to 20/80 in eleven weeks. Upon examining his eyes, his ophthalmologist told him that he had never seen that kind of regeneration before, and said, "Keep doing what you're doing. You may change some minds."

I realize that this book the outgrowth of the experience of only one man, my husband Mark, but given a choice between legal blindness if you live long enough and making nutritional changes, does it not seem that nutritional treatment is worth a try?

Keep reading for more about how to implement nutritional changes and additional ways to help yourself.

---

19 Crittenden, John. *Blind Faith.* (2014).

Principles for Beating Macular Degeneration

The last chapter discussed ways to beat macular degeneration which included an extremely nutritious diet, supplements and, in addition, nutritional IVs for more advanced cases. Since macular degeneration is starvation of the retina[1], excellent eye nutrition is the cornerstone of beating macular degeneration. Supplements are easy to take, but consuming a diet of the right high-nutrient foods may be even more important. There are also a number of damaging factors, influences and substances to avoid and lifestyle changes that can help you beat macular degeneration.

The **first and most important principle** for beating macular degeneration **is to eat the best possible diet for eye health**. This has become less easy to do over the last several decades as giant agribusinesses have replaced small family farms and the use of many agricultural chemicals has become the norm. Chemical fertilizers are very limited in the nutrients they contain and do not produce plants as high in trace minerals and other nutrients as grocery store produce contained decades ago.

Our meat and dairy products have also changed. During our great-grandparents' days, all animals were raised naturally and humanely. The chickens and turkeys pecked the ground looking for bugs to eat and the cattle grazed on grass. This yielded meat, eggs and dairy products with healthy ratios of omega 3 to omega 6 fatty acids.[2]

Artificial foods such as trans-fats did not exist before the 1940s. Highly processed and convenience foods were also products of the early and mid-20th century.

Wheat was intensively hybridized from the 1940s through the early 1980s to produce high-yielding wheat that is easier to harvest and thresh. This modern wheat is also higher in types of gluten that are more likely to cause celiac disease or gluten intolerance;[3] hence the epidemic of these problems. To read more about this, see http://www.healingbasics.life/how-wheat-changed.html.

Unsurprisingly, the incidence of food allergies has also risen dramatically in the last 25 years. Although the FDA did no testing before approving GMOs and insists that they are safe, the timeline associated with the epidemic of peanut anaphylaxis in children casts grave doubt on the safety of GMOs.[4] In 1996, GMO soy, the first

---

1 Abel, Robert, Jr. MD. *The Eye Care Revolution: Prevent and Reverse Common Vision Problems.* (New York, NY: Kensington Publishing Corp., 1999, 2014), 170

2 Servan-Schreiber, David, MD, PhD. *AntiCancer: A New Way of Life.* (New York: Penguin Group, Inc.,2009), 73.

3 Davis, William, MD. *Wheat Belly.* (New York, NY, Rodale, Inc., 2011), 25-26; also Gao, X; Liu, SW, et al. "High frequency of HMW-GS sequence variation through somatic hybridization between *Agropyron elongatum* and common wheat." *Planta* 2010 Jan;23(2)245-50.

4 O'Brien, Robyn. *The Unhealthy Truth: How Our Food Is Making Us Sick and What We Can Do About It.* (New York, Broadway Books, Random House, 2009), 65.

GMO food the FDA approved, came into widespread use. Soy and peanuts are both in the legume family, and it is not unusual for a person allergic to one member of a food family to react to other members as well. In 1997, the incidence of peanut anaphylaxis rose by 20%. This rate of increase in the United State has continued every year to the present time. When my children were elementary school students in the 1990s, none of their classmates suffered from peanut anaphylaxis. Now one in thirteen elementary school children has peanut anaphylaxis, meaning there are on average two to three sensitive students in each class.

In 1998, GMO soy was introduced in the United Kingdom, and the rate of soy allergies rose 50% in 1999.[5] Although these examples may not constitute proof that consuming GMOs is unsafe, I personally do not want to develop more food allergies, so I avoid GMO foods. The current list of GMO foods in the United States includes soy, corn, sugar beets (beet sugar), canola (oil), cottonseed (oil), papaya, zucchini, yellow summer squash, some tomatoes, some apples (including non-browning apples), some potatoes (but not sweet potatoes or yams) and alfalfa. When you shop for these foods fresh, if you want to avoid GMOs, you must buy organic produce. For packaged foods, look for a logo that says the food is certified GMO free.

This book offers help for coping with the changing dietary demands modern life and food have given to an increasing number of people. Aside from a few wheat-containing breads, the **recipes here are gluten-free and food-allergy friendly**. Therefore, readers who have these dietary challenges can **improve their eye nutrition while adhering to their special diets and thus helping keep inflammation levels in check, which is important for limiting the progression of macular degeneration.**

The **second principle** for beating macular degeneration is to **protect your eyes from high levels of blood glucose**,[6] also called high blood sugar. Both the retina and lens of the eyes are adversely affected by high or widely fluctuating blood glucose levels. The chapter beginning on page 40 and appendix sections beginning on pages 177 to 195 will help macular degeneration patients with type II diabetes, pre-diabetes, insulin resistance or reactive hypoglycemia to achieve stable blood glucose levels. Ophthalmologists Marc and Michael Rose report that the majority of their patients with type II diabetes can achieve control of their blood sugar using diet, supplements and exercise.[7] All the carbohydrate-containing recipes in this book give carbohydrate units per serving to assist diabetics and others with blood sugar problems in planning their diets. Some recipes also give protein units to help individuals following the insulin resistance diet to link and balance protein and carbohydrate units in meals.

The **third principle** for beating macular degeneration is to **protect your blood vessels**. The choroidal capillaries in the macula are very small and can eas-

---

5 O'Brien., 66, 89-90.

6 Rose, Marc R., MD and Michael R. Rose, MD. Save *Your Sight! Natural Ways to Prevent and Reverse Macular Degeneration.* (New York, NY: Warner Books, 1998),112.

7 Rose, 115-116.

ily become clogged.[8] Wet macular degeneration develops when they bleed or leak. The chapter beginning on page 47 gives nutritional and lifestyle advice to help keep these capillaries in good condition.

The **fourth principle** for preventing or beating macular degeneration is to **protect your eyes from damaging light**. This includes ultraviolet light (UVA and UVB) and, according current sources, high energy blue light. The chapter beginning on page 50 explains how light damages the retina and provides information on nutrients and sunglasses that prevent damage.

The **fifth principle** for beating macular degeneration is to **avoid drugs that may damage the eyes**. The chapter beginning on page 53 tells how various drugs cause eye problems and lists drugs to avoid.

The **sixth principle** for beating macular degeneration is to **avoid toxins and eliminate them from your body**. Avoidance is covered in detail as is drinking sufficient water, a practice that will help clear your body of toxins. Other practices which may help eliminate toxins are also discussed briefly in the chapter which begins on page 59.

The **seventh principle** for beating macular degeneration is to **exercise**. Strenuous exercise is not needed. Moderate exercise such as walking will help your eyes and, especially if done outdoors in a natural area, can be enjoyable. Read more about exercise starting on page 65.

The **eighth principle** for being kind to your eyes and to yourself in general is to **pursue peace and promote dominance of the parasympathetic nervous system** (the "rest and digest" part of the nervous system), as opposed to the "fight or flight" sympathetic nervous system. This information begins on page 67.

At first, following the principles above may seem overwhelming. Getting help can make implementing them easier and optimize the effects of the changes you make. It may be beneficial to consult a naturopath, an expert in nutritional health treatment. Mark's naturopath ordered blood tests and spent considerable time talking to him to determine what he needed and how to personalize the principles here to produce the best results. To find a naturopath nearby, see "Sources," page 221, for a website that lists naturopaths by location.

Following these principles will promote better health of your whole body as well as your eyes. You have everything to gain and nothing to lose.

---

8 Rose, 56.

Superior Eye Nutrition

Nowhere has there been so much scientific documentation about nutritional prevention and treatment as in the case of macular degeneration. The irony lies in the fact that this information is often virtually ignored by the very specialists who manage the patients with this disease.[1]

- Robert Abel Jr., MD

Our initial experience with eye doctors demonstrates the truth of this quote from ophthalmologist Robert Abel, MD. In contrast, at our first visit to the naturopath who is treating Mark, she said that she had worked with a medical doctor who gave nutritional IV therapy to macular degeneration patients and that it kept their macular degeneration from progressing. However, you may not need IVs if your macular degeneration is still in the early stages, and may even notice improvement without IVs from more advanced stages.[2] Highly nutritious food and supplements have great potential for helping your eyes and are less costly.

When we discussed Mark's diet and supplements on that first visit, the naturopath said, "Food is best." In food, vitamins and minerals come with other nutrients (some of which may have yet to be discovered) that synergistically help the body to best utilize all the nutrients present. Therefore, nutrients from food improve both eye health and general health most efficiently.

The nutrients in foods are classified as macronutrients or micronutrients. Macronutrients are present in large amounts and include protein, carbohydrate and fat. Micronutrients are found in small amounts and include vitamins, minerals, bioflavonoids, phytonutrients, essential fatty acids, antioxidants, carotenes, eye-protecting pigments such as lutein and zeaxanthin, and more.

Although the levels of micronutrients in foods have decreased due to modern agricultural methods, our food still contains plenty of protein, carbohydrate and fat. However, the quality of these macronutrients has deteriorated over the last several decades. Carbohydrate foods are usually highly refined and the fatty acid profile

1 Abel, Robert, Jr. MD. *The Eye Care Revolution: Prevent and Reverse Common Vision Problems.* (New York, NY: Kensington Publishing Corp., 1999, 2014), 170.

2 In *Blind Faith*, John Crittenden writes about how he progressed to legal blindness (vision **with glasses** of 20/200 or worse) while taking monthly injections of anti-VEGF drugs. (See page 29 for more about these drugs). He then began eating an extremely nutritious diet, making lifestyle changes and taking supplements, and after eleven weeks his vision had improved to 20/80. The Tahoma Clinic website contains a testimonial about reversing macular degeneration with supplements at the end of this page: http://tahomaclinic.com/treatment-for-dry-macular-degeneration/ The Tahoma Clinic does not claim that all patients have these results, but says that they help about 70% of macular degeneration patients. Therefore, consider refusing to accept a "no hope for improvement" prognosis without giving nutrition a try.

of feed-lot-fed cattle and chickens that spend their lives confined in small indoor spaces has become quite unbalanced.[3] The ideal solution for fatty acids is to eat meat, eggs and dairy products from free-range animals. Realistically, this may be too pricey for many of us. If so, eat lean cuts of meat and low-fat dairy products and make up for what is lacking in omega-3 fats by taking a high-quality omega-3 supplement daily. Be sure this supplement contains rosemary or other ingredients to prevent rancidity, keep it refrigerated, and use it in a reasonable amount of time. Rancid fats are damaging to blood vessels in the eyes.[4]

If you do not have an egg allergy, eggs from naturally raised chickens may be a high-quality protein option that is relatively inexpensive and also contains lutein and other nutrients that are especially good for macular degeneration. Our health food store has frequent long-lasting sales on eggs. Usually the sale price of these natural eggs is less than the non-sale price at the grocery store for pale-yolked, thin-shelled eggs of chickens raised conventionally.

Caveat: do not eat eggs in large quantities or every day because they are quite allergenic and it is easy to develop an egg allergy. Space out eating eye-healthy egg yolks at intervals of at least four or five days to help prevent allergies. However, you do not have to fear eggs because they contain cholesterol. See pages 47 to 48 for more about cholesterol, including that less than 10% of the cholesterol in our blood comes from our diet. Also discussed there is a study which showed that a low-fat low-cholesterol diet did not change the incidence of heart attacks. Avoidance of dietary fat and cholesterol has very little effect on blood cholesterol. However, if you avoid sugar and highly refined carbohydrates, you will improve your HDL/LDL cholesterol ratio, which is the only blood test result that actually correlates with improving blood vessel and heart health.[5]

Consider adding plant-source proteins such as legumes and nuts to your diet. I purchase many raw nuts from our local health food store for blood sugar stabilizing snacks. If you have difficulty digesting nuts and seeds, see pages 162 to 163 for recipes that neutralizes the phytic acid and enzyme inhibitors nuts contain and thus makes nuts easier to digest. To purchase easily-digested sprouted nuts and nut butters, see "Sources," pages 218 to 219. For easy meals, also see the legume soup recipes on pages 103 to 107.

Carbohydrate foods range from very good nutritionally to very unhealthy. **The carbohydrates in fruits and vegetables come with vitamins, minerals, and eye-healing substances such as lutein, zeaxanthin, carotenoids and anthocyanins. Many fruits and vegetables are low on the glycemic index so promote stable blood glucose levels** and can be eaten freely by diabetics, pre-diabetics and individuals with insulin resistance or who are following a gly-

---

3 Servan-Schreiber, David, MD, PhD. *AntiCancer: A New Way of Life.* (New York: Penguin Group, Inc.,2009), 73.

4 Rose, Marc R., MD and Michael R. Rose, MD. Save *Your Sight! Natural Ways to Prevent and Reverse Macular Degeneration.* (New York, NY: Warner Books, 1998), 29.

5 Thompson, Rob, MD and Dana Carpender. *The Glycemic Load Diabetes Solution.* (New York, NY: McGraw-Hill, 2012), 48.

cemic-control weight loss plan. The vegetables which are highest in the nutrients needed for macular degeneration are dark green leafy vegetables such as kale, spinach, collards, parsley and other dark green vegetables and orange, yellow, red and purple vegetables such as squash, red, orange and yellow peppers, tomatoes, purple cabbage, sweet potatoes, etc. Blueberries are a great source of anthocyanins. Shop the produce department for *color!*

Starchy carbohydrate foods can be problematic. Most processed foods are made with highly refined white flour, corn sweeteners and/or sugar. Ophthalmologists Marc and Michael Rose, MD, say, "Processed foods and sugar are anti-nutrients because they actually drain nutrients instead of contributing to them."[6] Your body has to use stored nutrients to digest and metabolize theses foods. For the sake of their eyes, **macular degeneration patients should avoid all processed foods.**

Grains supply vitamin E, B vitamins, fiber and other nutrients. However, foods made from them are often highly refined and are usually high on the glycemic index scale. The naturopath told Mark, "No white flour, white rice, white potatoes or sugar." Eat whole grains for their nutrients but, if you eat wheat, make sure your whole wheat bread contains more than just a small amount of whole wheat. Most commercially made whole wheat bread contains mostly white flour, a little finely ground whole wheat flour and colored ingredients to make the bread look more brown. Most whole wheat flour is as finely ground as white flour and therefore just as high-glycemic as white flour. Stone ground whole wheat flour makes bread with a lower glycemic index score. (See the 100% stone ground whole wheat bread recipes on pages 131 to 133). The best choices among grain foods are items baked using stone-ground flour such as Bob's Red Mill stone-ground quinoa flour, 100% bran cereal, 100% stone-ground whole wheat bread and traditionally-made long-fermented sourdough bread in which the acid lowers the glycemic index score of the bread. Some breads called sourdough have not undergone a long fermentation but have had tart-tasting ingredients and flavorings added instead. Therefore, ask questions about how a brand of sourdough bread is made before you purchase it or visit the website of the makers of sourdough bread to be sure it has undergone a traditional long sourdough fermentation. See pages 139 to 144 for sourdough bread recipes made more easily with a freeze-dried *Lactobacillus* starter using a programmable bread machine or by hand. In addition to whole wheat sourdough bread recipes, this book includes recipes for wheat-free and gluten-free sourdough bread.

Starchy vegetables include corn, white potatoes and sweet potatoes. Baked white potatoes have a glycemic index score of 111 which is greater than that of pure glucose (100). They possibly should be eaten only by individuals whose blood sugar is very stable, and they should always be eaten with a balancing serving of protein and some fat to slow the absorption of the carbohydrates. Baking makes the glycemic index of all potatoes rise, both white potatoes and sweet potatoes.

Sweet potatoes are high in eye-supporting nutrients and are lower on the glycemic index than white potatoes. Boiled Jersey sweet potatoes, also called white sweet

---

6 Rose, 27.

potatoes although their flesh is yellow, have a glycemic index score in the low range (44) and are delicious mashed. See the recipe on page 115.

Corn has a low glycemic index score and is a nice treat when you want a change from dark green leafy vegetables. Yellow corn also contains lutein. Non-organic corn is usually a genetically modified organism (GMO) in the United States. You may have to search far and wide or grow your own corn to get yellow corn rather than highly hybridized, super-sweet white or bicolor corn. This year I planted heirloom organic corn seeds.[7] The corn was not candy-sweet like modern hybrid corn and Mark experienced no blood sugar destabilization when he ate it. He had not eaten corn for several years because purchased corn always put him "into orbit."

Avoid cold breakfast cereals except for 100% wheat bran. Other cold cereals are pricey and nearly devoid of nutrients since they are almost totally composed of highly refined flours and sugar. Instead try whole grains such as thick rolled oats, which have a lower glycemic index score than regular or even old-fashioned rolled oats. Unfortunately, instant oatmeal is highly refined, has a high glycemic index score, and should be avoided.

An ideal breakfast for diabetics should include enough protein to satisfy hunger, such as an omelet or Mark's usual breakfast protein which is eight ounces of low-fat cottage cheese. If desired, diabetics can have up to 30 grams of low glycemic-index carbohydrates such as fresh fruit or a whole grain product with an adequate serving of protein. See pages 45 to 46 and 179 to 180 for how to balance carbohydrates and protein in meals and snacks if you are diabetic, pre-diabetic or practice glycemic control to lose weight.

Healthy fats and cholesterol are important for good health. Some types of fat are essential, meaning that we can't make them but need them so we must ingest them. Cholesterol is also vital for health. We can and do make at least 90% of the cholesterol in our blood. Less than 10% of the cholesterol in our blood comes from our diets. Cholesterol is needed for making hormones, including sex hormones and some adrenal hormones. It is crucial for eye health because it is needed for transporting nutrients, including antioxidants, to the eyes as well as to other organs.[8] It also makes up part of the cell membrane of every cell in the body.

Rather than eating a very low fat (and thus probably a high carbohydrate) diet and taking drugs to lower cholesterol (statins),[9] we should make an effort to increase our "good" HDL cholesterol by exercising, by eating a low-carbohydrate diet and by consuming omega-3 fatty acids, olive oil, coconut oil, fatty fish, and purple vegetables and fruits.[10] HDL helps remove deposits from blood vessels in eyes, the heart, and all over the body.

---

7 Turtle Tree Seed sells seeds for the yellow corn of yesteryear, a heirloom variety called Ashworth, here: http://turtletreeseed.org/product/843-ashworth-sweet-corn/

8 Rose, 31.

9 Visit this page to learn more about why to avoid statins http://www.healingbasics.life/statins.html

10 Rose, 32 and Spritzler, Francesca, RD. "Nine Ways to Increase Your HDL Choloesterol Levels," https://www.medicalnewstoday.com/articles/318598.php

Healthy types of fat are your friend, not the weight-packing enemies you might have thought. The obesity epidemic began when Americans embraced low-fat and therefore high-carbohydrate diets. If we nearly eliminate one category of food, such as fat, we have to add food from another category to get enough food to be satisfied. These added foods are often refined carbohydrates such as processed foods made with highly refined white flour, sugar and corn sweeteners.

The type of fat eaten is very important for eye health. **Avoid all trans fats and rancid fats.** Trans fats are unnaturally saturated fats made by adding hydrogen to double bonds of oils. Foods that contain trans fats include most types of margarine, vegetable shortening, and many processed foods. Polyunsaturated fats such as corn, soy and safflower oil should also be avoided because they become rancid easily. Rancidity usually starts during their production, especially if they are not cold-pressed, so a just-purchased bottle of vegetable oil probably contains some rancid (damaged) fat as soon as you open it although you may not smell it. It will oxidize even more quickly after it is opened and become especially damaged if heated. Ophthalmologists Marc and Michael Rose, MD say, "Rancid oils directly contribute to the clogging of small blood vessels, including those in the eye,"[11]

Avoid deep-fat fried foods in restaurants because, even if they use a healthy type of oil, they re-use their frying oil. With repeated heating, the oil becomes more and more oxidized. I worked in the grill of a discount store when I was a college student and remember that the oil in the deep-fat fryer was strained every few days but it was discarded and replaced only about once every two or three months. The longer oil is heated, the greater the percentage of damaged fat it contains.

The Drs. Rose recommend using monounsaturated oils including olive, canola and avocado oil **because monounsaturated oils are the least likely to become rancid.**[12] Canola oil is controversial in the health-food community. Some bad press is justified because 80% of canola oil made in the United States is made from GMO canola seeds. However, organic non-GMO canola oil is readily available. If you wish to avoid canola oil due to the controversy and can eat olive and/or avocado oil continually without developing an allergy, feel free to do so. However, individuals who follow a rotation diet for food allergies must eat each food, including each kind of oil, only on every fourth day or at a longer interval. Therefore, they need more oil choices. Keeping food allergies and the inflammation they cause under control is essential for macular degeneration patients. To gain a balanced view of the controversy, learn about non-GMO organic canola oil and, in the first source, read about health studies on canola oil, visit these webpages: https://www.huffingtonpost.com/entry/can-the-right-kind-of-canola-oil-be-a-health-food_us_59775d13e4b01cf1c4bb7366 ; http://www.berkeleywellness.com/healthy-eating/food-safety/article/canola-oil-myths-and-truths ; https://www.prevention.com/food-nutrition/a20448772/is-canola-oil-safe/[13]

---

11 Rose, 29.

12 Rose, 30.

13 Christianson, Dr. Alan, ND. "Can the Right Kind of Canola Oil be a Health Food?" *HuffPost*,

Naturally saturated fats such as coconut and palm oil are also safe for your eyes and are best for cooking with high heat. Coconut oil promotes weight loss. Butter is also a healthy fat, especially if made from vitamin A-rich milk from grass-fed cows.

Have some healthy fat with every meal. Not only will you stay satisfied longer, but fat is essential for good absorption of fat-soluble vitamins and other nutrients including lycopene. Tomatoes are high in lycopene, a carotenoid phytonutrient that combats macular degeneration as an antioxidant, is anti-inflammatory and promotes blood vessel health. However, lycopene is not well absorbed and utilized if tomatoes are eaten raw or without fat.[14] Therefore, enjoy cooked tomatoes as part of a fat-containing meal as often as possible. At Mark's first visit with the naturopath, we discussed tomatoes. She was very pleased to hear that he eats Italian food every Friday night because the cooked tomatoes in a fat-containing meal give him lycopene.

According to ophthalmologist Michael Samuel, MD, research has shown that inflammation contributes to the development of dry macular degeneration. Therefore, he says it is important to **eat an anti-inflammatory diet.** This means that individuals with food allergies or gluten intolerance must avoid their problem foods. Dietary fat should contain less omega-6 fatty acids and more omega-3 fatty acids, monounsaturated fats, and naturally saturated fats. Most of the carbohydrates we eat should be low on the glycemic index to keep from fueling inflammation. Also, try to include generous amounts of anti-inflammatory foods in your diet. See the list of such foods on pages 196 to 197.

About eight months after the original publication of this book, we discovered **omega-6 pro-inflammatory fats in a surprising place – the SALMON that most macular degeneration patients eat weekly** or more often! Here's how it happened: Mark began getting hives on his abdomen every time he ate salmon so stopped eating it. When he had a regular checkup with the retina specialist, she insisted, and rightly so, that he get back to eating oily fish once a week. When I told a friend of mine about this, she said that she reacted to fresh salmon but not to canned wild-caught salmon. Since Mark dislikes canned fish, I bought fresh wild-caught salmon, and he has eaten it weekly with no problems. We searched online to get insight on his reactions and discovered that farmed salmon are fed antibiotics among other awful things. Mark is allergic to penicillin and my friend has multiple antibiotic allergies, which explains their reactions.

October 2, 2017. https://www.huffingtonpost.com/entry/can-the-right-kind-of-canola-oil-be-a-health-food_us_59775d13e4b01cf1c4bb7366 ; "Canola Oil Myths and Truths," *Berkeley Wellness*, February 25, 2017. http://www.berkeleywellness.com/healthy-eating/food-safety/article/canola-oil-myths-and-truths ; Ecklekamp, Stephanie. "Is Canola Oil the Toxic Bomb You've Been Led to Believe It Is?" *Prevention.* March 16, 2015. https://www.prevention.com/food-nutrition/a20448772/is-canola-oil-safe/ . Also see Broaddus, Hannah. "Debunking 5 Myths About Organic Canola Oil." June 5, 2017; http://www.centrafoods.com/blog/debunking-5-myths-about-organic-cavnola-oil

14 Axe, Josh, DC. Lycopene: A Powerful Antioxidant to Help Prevent Cancer & Keep Your Heart Healthy. https://draxe.com/lycopene/

The krill that wild salmon eat are an excellent source of omega-3 fats. **The poor diet farmed salmon are fed results in a high level of omega-6 fatty acids in their bodies.** This is not even half of the bad news about what farmed salmon are fed. Read the rest of the story here: https://www.tampabay.com/things-to-do/food/cooking/The-facts-about-farmed-salmon-you-wish-you-didn-t-know_166193900 **For best eye health, eat only wild-caught salmon, fresh, frozen or canned.** Canned salmon is economical and can be delicious as "Salmon Burgers," page 100.

The macronutrients are easy to consume in sufficient quantities. Micronutrients can be more elusive. The best way to get the micronutrients your body needs is to consume as much organic food as you can afford. See page 34 for advice on how to get enough protein without spending much on naturally raised meat. Organic produce is only a little more costly than conventional produce, and fruits and vegetables are the best sources of many micronutrients, so purchasing organic produce is wise use of your dollars. Watch for sales on organic produce which can cost less on sale than conventional produce.

In addition to consuming the healthy foods mentioned in this chapter, you must be able to absorb them. "You are what you eat" isn't always true. You must adequately digest your food to absorb the nutrients it contains. Many adults over age 60 are unable so secrete sufficient hydrochloric acid (HCl) for good digestion. This lowers absorption of B vitamins, zinc and other minerals. There are remedies for this problem such as bitters or apple cider vinegar with meals that sometimes help.[15] However, considering the importance of your eyes, if you need more help, you may wish to consider HCl supplements. Consult a holistic health practitioner for how much and how to take HCl with meals. For more information about taking HCl supplements, see pages 55 to 56 or the last three paragraphs of this webpage: http://www.healingbasics.life/proton-pump-inhibitors.html

The bottom line on ideal nutrition for your eyes is to avoid foods that come in packages and instead eat whole natural foods, especially including fatty fish, eggs (four yolks per week are recommended for macular degeneration), kale, spinach and other dark green leafy vegetables, and orange, red, yellow and purple produce. You will grow to enjoy all the bright flavors of these colorful foods. Bon appetit!

---

15 Rose, 142.

Protect Your Eyes from High Blood Glucose

Good blood glucose control is essential for eye health. Both the retina and lens are adversely affected by high blood glucose (also called high blood sugar). High blood glucose, whether constant or intermittent, causes macular edema (swelling) and also contributes to the development of macular degeneration and cataracts.

Individuals with diabetes, pre-diabetes or insulin resistance, in my opinion, need to take control of their diets to prevent losing their sight. Even "normal" non-diabetics who have macular degeneration also would do well to strive to keep their blood glucose levels stable.

Disclaimer: The material about diabetes below and elsewhere in this book applies only to adult-onset type II diabetes. Consult your doctor for specific advice, meant especially for you, about how you should manage your diet. The information here is only my opinion about what is helpful for one person, my husband Mark.

This chapter discusses several dietary programs used by diabetics in their quest to achieve good control of their blood sugar. On pages 43 to 46 is "bottom line" information which is what seemed most helpful for Mark. Keep in mind that we are all unique, and what works for one person may be different from what is the best way for another person to achieve good control of blood glucose levels.

When Mark heard that his hemoglobin A1c blood test indicated pre-diabetes, the question in my mind was, "How does he really get control of his blood sugar level?" He had been following (but not with absolute strictness) the insulin resistance diet[1] for about eight years. (See details about the insulin resistance diet on pages 178 to 180). It probably helped keep him from developing full-blown diabetes, but I thought we needed to do better. I turned to the internet for dietary advice for diabetics and was shocked by what I read! The American Diabetes Association (ADA) promotes diets that are much more liberal with carbohydrates than what he had been eating. I read an ADA recommendation that large people with active lifestyles could eat up to 75 grams of carbohydrate per meal, two and a half times as much as the insulin resistance diet allows in any two-hour window. The reason 30 grams of carbohydrate is the most allowed in two hours is because if we eat more, our bodies will enter the "deposit this food as fat" mode resulting in weight gain. A weight increase is detrimental because a BMI (body mass index) in the overweight or obese range is associated with insulin resistance and increased diabetes risk.

Mark told me about a co-worker who has adult-onset diabetes which is so severe that he takes insulin. This man told Mark that his wife makes him two pans of sticky buns per week, "but it's OK, because I balance it with medication." Mark's coworker

---

1 Hart, Cheryle R., MD and Mary Kay Grossman, RD. *The Insulin Resistance Diet.* (New York, NY: McGraw-Hill, 2008).

follows the ADA diet given to him by his physician. The ADA's acceptable range for fasting blood sugar tests goes up as high as 130 mg/dl. (Normal fasting blood sugar is up to 99 or 100 mg/dl). The normal value for blood sugar two hours after meals is up to 140 mg/dl. The ADA says readings of up to 180 mg/dl two hours after meals are acceptable. They call this "controlled" diabetes.

The National Institute of Diabetes and Digestive and Kidney Diseases says, "Even when diabetes is controlled, the disease can lead to chronic kidney disease and kidney failure."[2] In other words, serious complications are expected for diabetics who "control" their diabetes to ADA standards using carbohydrate-liberal diets which they "balance" with medication.

Conventional doctors have their patients monitor only their fasting blood sugar, meaning that the test is taken before breakfast. They are instructed to not test their blood sugar after meals. However, ophthalmologists Marc and Michael Rose, MD, encourage diabetic patients to test their blood sugar two hours after meals to determine what foods may be safely eaten and in what quantities.[3] It seems to me that the ideal diabetic diet should be what works best for the individual, and using glucometer readings after meals to discover what keeps blood sugar relatively stable makes good sense.

Dr. Rob Thompson, a cardiologist who developed diabetes, used his own response to diet as measured by a glucometer to work out a dietary system that has not only helped him but many of his patients and readers of his books. It is a "middle of the road" system between the ADA diets described above and on the previous page and the low-carbohydrate high-fat (LCHF) diet described on the next page.

Dr. Thompson says that diabetic damage to blood vessels takes years to manifest itself. The diabetic patient seems to be doing all right with lax dietary standards and blood sugar lowering medications which reduce blood sugar enough to produce fasting blood sugar test results which the ADA says are acceptable (up to 130 mg/dl). He says that by the time pre-diabetes develops, the individual is having blood sugar surges that can damage blood vessels after starchy meals. Therefore, he advises stricter control and limits his carbohydrates from starches to 30 grams per day. In addition, he eats as much as he wants of fruits and non-starchy vegetables, most of which have a low glycemic index score. Being only human, Dr. Thompson indulges his sweet tooth with a little dark chocolate or other sweets after dinner most evenings and eats pasta but takes the drug Acarbose™ on pasta nights. This is a drug that slows carbohydrate absorption. However, he says that "a very starchy meal can overwhelm its effects and cause an undesirable blood sugar surge.[4]

At the strict end of the diabetic diet discipline spectrum is Dana Carpender, author of *The Low-Carb Diabetes Solution Cookbook,* and her colleagues at the

---

2 Carpender, Dana. *The Low-Carb Diabetes Solution Cookbook.* F(Beverly, MA: Fair Winds Press, 2016), 14.

3 Rose, Marc R., MD and Michael R. Rose, MD. Save *Your Sight! Natural Ways to Prevent and Reverse Macular Degeneration.* (New York, NY: Warner Books, 1998), 116.

4 Thompson, Rob, MD and Dana Carpender. *The Glycemic Load Diabetes Solution.* (New York, NY: McGraw-Hill, 2012), 128.

HEAL (Healthier Eating And Living) Diabetes and Medical Weight Loss Clinics. She not only teaches but personally follows a low-carbohydrate high-fat (LCHF) diet which includes only 20 grams of carbohydrate per day coming from non-starchy vegetables and salads. This diet is a ketogenic (ketone producing) diet, which means that the body burns fat for energy rather than glucose. Diabetics must be under medical supervision when they start the LCHF diet.[5] This is because the diabetic's blood glucose level must be monitored and glucose-controlling medication must be incrementally adjusted to "fit" the changes in metabolism. If this is not done, the patient can suffer severe low blood sugar or even insulin shock.

Using the LCHF diet, 95% of diabetic HEAL patients achieve normal fasting blood sugar (100 mg/dl or less), eliminate blood sugar surges, and avoid the risk of diabetic eye disease, amputations, kidney disease, etc. As many as 75% of patients eliminate all diabetes drugs and 20% reduce their need for medication. For additional information about the HEAL program, see this website: https://healclinics.com/

HEAL Director Eric C. Westman, MD, says that research has demonstrated the safety of eating more fat and very little carbohydrate.[6] Once the dieter has adjusted to the HFLC diet, the diet is easy to follow because the ketones produced by the body suppress appetite.[7] The results of blood tests for patients on this diet are impressive, such as reducing high triglycerides by 900 mg/dl or raising HDL by 50 mg/dl.[8] Dieters lose weight easily and without hunger while eating 2600 calories per day.[9]

However, Dr. Cheryle Hart, MD, author of *The Insulin Resistance Diet,* offers a different view about very low carbohydrate diets.[10] Her major concern is that fruits and whole grains, which are the best sources of many antioxidants and other vital nutrients, are eliminated or nearly eliminated on such diets. She says, "Lack of the nutrients they provide can reduce your ability to resist infection, cancer and aging."

She also notes that extremely low carbohydrate diets are usually low in fiber, and a high fiber diet can prevent diabetes as shown in several studies.[11] Her final concern is that it may be quite difficult to live with such a restrictive diet for the long term.[12] This may cause abandonment of any dietary control.

Our experience with making Mark's diet slightly lower in carbohydrates demonstrated the difficulty of following a very low carbohydrate diet. A few weeks after Mark decided his breakfast was too much food and gave up having a slice of sour-

---

5 Carpender, 17.

6 Carpender, 6.

7 Carpender, 16.

8 Carpender, 7.

9 Carpender, 17.

10 Hart, Cheryle R., MD and Mary Kay Grossman, RD. The Insulin Resistance Diet. (New York, NY: McGraw-Hill, 2008), 75.

11 Hart, 75-76.

12 Hart, 76.

dough whole wheat toast every morning, he began experiencing intense cravings for bread. We began adding a slice of sourdough whole wheat bread (recipe on page 139) to dinners of meat or fish, vegetables and salad once or twice a week. This has eliminated the cravings. Perhaps he was craving nutrients that are found in whole wheat bread.

Bottom Line Information on Blood Sugar Control

In *Save Your Sight!* Drs. Marc and Michael Rose, MD, do not give specific detailed diet advice for diabetics with eye disease but encourage general moderation, a diet with plenty of protein from eggs, meat, fish, poultry and tofu, abundant vegetables, some fruit and complex whole-grain carbohydrates, but the **elimination of all refined grains, sugar and processed foods.** They also recommend that **diabetics test their blood sugar four times a day** with a glucometer while they are adjusting to dietary changes to be sure that they are not having major fluctuations in blood sugar levels. They advise keeping a journal for three months of everything eaten or drunk, the day's events, blood sugar test results, exercise, etc. so the patient can know what raises their blood sugar, lowers it or keeps it stable. The information in the journal helps with adjusting diet, exercise, and other controllable factors to promote stable blood sugar. They stress that regular exercise is essential to good blood sugar control.[13]

What is eaten and drunk, when, and what accompanying foods are consumed at the same time can affect how much a meal impacts blood sugar levels. Some foods will reduce the glycemic impact of accompanying foods eaten at the same meal. Here are ways that you can reduce the glycemic impact of carbohydrate foods in a meal and keep blood sugar more stable:

1. **Never eat high carbohydrate foods alone or as the first bite of a meal.**[14] Eat at least a few bites of a protein food or high fiber foods such as a salad or fresh fruits or vegetables as the first bites of a meal before you eat any starchy or concentrated carbohydrate food. Eating protein and carbohydrate foods at the same time, such as in a sandwich, is also good.

2. **Fat slows stomach emptying so it will reduce the glycemic impact of a meal** and should be included in meals containing concentrated carbohydrates. Eating a handful of nuts, a few bites of cheese, or several olives 10 to 15 minutes before a meal will decrease the amount blood sugar rises after the meal.[15]

3. **Have a good helping of vinegar with your meal**, such as oil and vinegar salad dressing. (See the recipe on page 166). Acetic acid in vinegar inhibits the enzymes that digest starch, which slows absorption of glucose from the stomach.[16]

---

13 Rose, 116-117.

14 Thompson, Rob, MD and Dana Carpender. *The Glycemic Load Diabetes Solution.* (New York, NY: McGraw-Hill, 2012), 125.

15 Thompson and Carpender, 125.

16 Thompson and Carpender, 125.

4. **Be careful about what you drink other than water**. Do not drink carbohydrates. Dissolved sugar in beverages is absorbed much more quickly than the same amount of sugar in candy. Sodas containing high fructose corn syrup are the worst offenders because they contain instantly absorbable single sugars, glucose and fructose. Table sugar (sucrose) must be broken into two molecules, glucose and fructose, before it can be absorbed.[17] Fructose impacts blood sugar levels more slowly than glucose because it takes time for it to be converted to glucose by the liver before it is sent to the rest of the body to be used.

Although **fruit juice** contains vitamins and other nutrients, the natural sugars it contains are absorbed very rapidly, making juice likely to cause a blood sugar surge similar to the surges caused by soda.[18]

Non-sugary beverages may or may not cause problems. **Milk** does not raise blood sugar levels in most people and is a better choice than soda or fruit juice. However, some diabetics report that milk does raise their blood sugar moderately. If you are a diabetic, test your blood sugar two hours after drinking a glass of milk alone, without other foods, to determine whether it is all right for you.[19]

Alcoholic beverages such as wine and liquor theoretically should be acceptable because the sugar they began with has been fermented to alcohol. Alcohol does not satisfy hunger, but it supplies calories so should be consumed moderately to control weight. Beer does contain some carbohydrate so should be limited to 12 ounces and be consumed with protein-containing food. Mixed drinks, if made with sugar or other carbohydrate-containing ingredients, should be avoided.[20] Individuals who are allergic to brewer's yeast should avoid all alcoholic beverages. I have heard about non-diabetics with food allergies having blood sugar test results as high as 250 mg/dl two hours after consuming an allergy-problem food.[21]

Although **diet sodas** are usually considered "safe" for blood sugar control because they do not contain carbohydrates, there is evidence that artificial sweeteners contribute to weight gain and insulin resistance. (Insulin resistance is the root problem in metabolic syndrome which can be a precursor of diabetes. For more about this, see page 177). If you are inclined to drink diet soda, please read the first online reference, especially, and the others footnoted below.[22]

---

17 Thompson and Carpender, 66-67.

18 Thompson and Carpender, 69-70.

19 Thompson and Carpender, 72.

20 Thompson and Carpender, 72.

21 Information from Katherine Gibbons, PhD, nutritionist and founder of Healthy Actions, LLC.

22 Sandou, Ana. "Diet Soda Sweetener May Cause Weight Gain," *Medical News Today*. November 25, 2016. https://www.medicalnewstoday.com/articles/314345.php ; Gul, Sarah Shireen et al. "Inhibition of the gut enzyme intestinal alkaline phosphatase may explain how aspartame promotes glucose intolerance and obesity in mice." *Applied Physiology, Nutrition and Metabolism*, doi: 10.1139/apnm-2016-0346. http://www.nrcresearchpress.com/doi/10.1139/apnm-2016-0346#.W0ZsEtJKiM-, Fowler, SP et al. "Fueling the obesity epidemic? Artificially sweetened beverage use and long-term weight gain." *Obesity*. 2008 Aug;16(8):1894-900. https://www.ncbi.nlm.nih.gov/pubmed/18535548 and Stellman, SD, Garfinkle, L." Artificial sweetener use and one-year weight change among women." *Preventive Medicine*. 1986 Mar;15(2):195-202. https://www.ncbi.nlm.nih.gov/pubmed/3714671

When we heard concerns about artificial sweeteners promoting weight gain and insulin resistance, we tested Mark's blood sugar after a meal of steak, non-starchy vegetables (purple cabbage), salad with oil and vinegar dressing, and a caffeine-free diet Coke™. His two-hour postprandial blood sugar was 98 mg/dl. Although the meal contained almost no carbohydrate from the vegetable and salad, the test result was the same as for a previously tested meal which included half of a large sourdough pizza. The impact of the soda on his blood sugar was clear to him. I had told him that I read in *Save Your Sight!* that aspartame was an eye toxin. That was the last diet soda he had.

For more about aspartame being toxic to eyes, see pages 59 to 60. For a cola recipe sweetened with stevia or monk fruit, see page 91.

Caffeinated beverages are usually considered safe for blood sugar control, but they should not be consumed without food. In *The Insulin Resistance Diet,* Cheryle Hart, MD, writes that caffeine, nicotine and artificial sweeteners raise insulin levels. Caffeinated beverages such as unsweetened coffee and tea do not raise blood glucose levels but, because they raise insulin, blood sugar can drop and lead to cravings, hunger and fluctuations in blood glucose levels. However, caffeinated beverages may be enjoyed safely with a protein-containing meal or snack.[23] To prevent forgetting and drinking regular coffee or tea without food, consider switching to decaffeinated varieties.

6. **What is eaten early in the day influences the glycemic impact of meals eaten later in the day. Have a high-protein blood sugar stabilizing breakfast.** Eating a high starch, or even worse, a high starch plus sugar breakfast (such as most cold cereals) will not only cause a blood sugar surge after breakfast, but will make a moderate lunch cause a larger surge than the same lunch after a high-protein breakfast.[24] A high carbohydrate breakfast will also make you hungrier later in the day. A study found that those who ate a high carbohydrate breakfast consumed 145 more calories for lunch than when they ate a high protein breakfast.[25]

A few principles from *The Insulin Resistance Diet* by Cheryle Hart, MD, add some final bottom line advice for controlling blood sugar.

7. **Choose most carbohydrate containing foods from those with a low glycemic index score.** Eat medium GI foods only occasionally and high GI foods rarely. To see tables of GI scores for foods, turn to pages 183 to 195.

8. **Balance protein and carbohydrate in every meal, limiting the amount of carbohydrate in any meal to 30 grams (2 carbohydrate units).** If you eat two units of carbohydrate for a meal, you must eat at least two units of protein (14 grams) to balance the carbohydrate. You can eat as much additional protein as needed to satisfy hunger. Round out the meal with non-starchy vegetables. A little healthy fat, such as monounsaturated olive oil in a salad dressing, is an especially good addition to meals or snacks containing low-fat protein.

---

23 Hart and Grossman, 22.

24 Thompson, Rob, MD. *The Glycemic Load Diet.* (New York, NY: McGraw-Hill, 2006), 61.

25 Thompson, 62.

The time window for eating 30 grams of carbohydrate is two hours. Eat no more than 30 grams of carbohydrate in any two hour period or the body will store the excess carbohydrate as fat. Planning for and knowing that you will be having a protein-containing snack in two to three hours should make keeping meals moderate in carbohydrates and size a realistic goal.[26]

9. **Eat at least three protein containing snacks per day at mid-morning, mid-afternoon and in the evening or at bedtime**.[27] They don't have to be large; a handful of nuts will do. If you haven't eaten carbohydrate foods in the last two hours, occasionally you may add carbohydrates to your snack in the correct balance with the protein you are eating. Eat enough at each snack to quell hunger. If you are hungry an hour or two after a meal or snack, an additional protein snack is all right, but do not eat more carbohydrate and exceed 30 grams in two hours.

10. **Eat breakfast early, ideally within the first hour after arising**. Your breakfast should contain plenty of protein. If you want carbohydrate, this is the best meal to limit your carbohydrate consumption to one unit (15 grams) to decrease hunger later in the day. (Please pardon the repetition that follows, but...) The maximum amount of carbohydrate allowed with breakfast or any meal or snack is 30 grams. Balance two units (30 grams) of carbohydrate with two or more units of protein (14 grams *or more*), or enough protein to satisfy hunger. Most carbohydrates eaten at breakfast and for the rest of the day should have low glycemic index values.

For more about the science behind the insulin resistance diet and specific details about the diet, see pages 177 to 180. For tables of foods with their glycemic index scores and the serving size which is one carbohydrate unit, see pages 183 to 195.

Controlling Mark's blood sugar with enough strictness to avoid eye-damaging blood sugar surges but with some leeway for enjoying food was our goal. How to achieve this is an individual process. If you're diabetic or pre-diabetic, testing blood sugar two hours after meals will help you make decisions about what foods to eat and in what quantities and combinations. Your blood sugar two hours after a meal should be 140 mg/dl or less. In Mark's case, glucometer use revealed that he does not have two-hour after meal blood sugar levels over 140 mg/dl or even near this level.[28]

Pay attention to flavors, textures and aromas while eating and enjoy your food as you use the information in this chapter to achieve good blood glucose control.

---

26 Hart and Grossman, 33.

27 Hart and Grossman, 82

28 When we began using a glucometer to test Mark's blood sugar two hours after meals, we discovered that he never had results higher than the upper limit of normal (140 mg/dl) for two hours after meals, even when he ate a meal with as much pasta as he wanted containing 60 grams of carbohydrate or a restaurant meal containing a "normal" (not sourdough) white flour hamburger bun and a large portion of fries. See pages 19 to 20 for more about this and how the hemoglobin A1c blood test can be misleading.

Protect Your Blood Vessels

All blood vessels in the same body are subjected to the same stresses, but the choroidal capillaries which nourish the macula are more at risk for damage because they are very small and fragile. When we implement nutritional and lifestyle changes which are beneficial for the arteries of the heart, the blood vessels in the retina and macula also benefit.

The conventional medical system's Holy Grail for blood vessel health is lowering patients' cholesterol blood tests results. We are bombarded with commercials for cholesterol lowering drugs (statins) every time the television is on. In *Save Your Sight!* ophthalmologists Marc and Michael Rose, MD, point out that the studies that indicate that cholesterol and blood pressure must be lowered to prevent or treat heart disease are funded by pharmaceutical companies.[1] They also point out that in spite of the high percentage of Americans taking both blood pressure and statin drugs (which are very effective for lowering blood pressure and cholesterol blood test numbers), the incidence of heart disease has not declined.[2]

Low-fat, low-cholesterol diets also do not help blood vessel-related heath problems. The large Women's Health Intensive Dietary Modification Trial (2006) investigated the effects of low-fat low-cholesterol diets on the incidence of heart disease. Although the women in the low-fat low-cholesterol diet group successfully lowered their fat and cholesterol consumption, their cholesterol blood test results decreased only 2% and there was no difference in the incidence of heart attacks between the diet group and the control group of women.[3]

A study done in 2009 analyzed data from thirty studies on the effects of low-fat diets versus moderate-fat diets on the risk of heart attacks. The data analysis showed that low-fat, low-cholesterol diets did not lower the amount of "bad" LDL cholesterol in the blood but did reduce the amount of "good" HDL cholesterol which worsened the balance between good and bad cholesterol. The HDL/LDL ratio is the only cholesterol blood test number which is a good predictor of heart or artery disease.[4] Therefore, the study results suggest that low-fat, low-cholesterol diets might be increasing the risk of heart and artery disease.

The practical question for those with macular degeneration is, "What dietary changes *will* help prevent damage to blood vessels?" In *Save Your Sight!* ophthalmologists Marc and Michael Rose say that the most important fat-related changes to make are:

(1) Make the **major source of fat in your diet stable monounsaturated oils** such as olive, avocado and organic non-GMO canola oil. (See pages 37 to 38 for

---

1 Rose, Marc R., MD and Michael R. Rose, MD. Save *Your Sight! Natural Ways to Prevent and Reverse Macular Degeneration.* (New York, NY: Warner Books, 1998), 167.

2 Rose, 167.

3 Thompson, Rob, MD and Dana Carpender. *The Glycemic Load Diabetes Solution.* (New York, NY: McGraw-Hill, 2012), 47.

4 Thompson and Carpender, 48.

information about canola oil controversy and safety). Monounsaturated oils are the least likely to become rancid (damaged) so are safest for the delicate blood vessels of the eye. Ophthalmologists Marc and Michael Rose, MD say, "Rancid oils directly contribute to the clogging of small blood vessels, including those in the eye,"[5]

(2) **Naturally saturated fats** such as coconut and palm oils and butter may be **consumed in moderation** and are best for high temperature cooking.

(3) **Polyunsaturated fats should be avoided** because they become damaged (rancid) easily. Fats that are polyunsaturated include corn, soy and safflower oil.

(4) **Trans-fats** are unnatural harmful fats so **should be avoided completely**. Trans fat foods include most margarines and vegetable shortening.[6]

An additional dietary change they recommend is to **totally avoid sugar and refined white flour products**[7] **because the incidence of artery disease is highly correlated with sugar consumption, not cholesterol consumption.**[8]

The best dietary way to optimize the triglyceride level and HDL/LDL ratio is to achieve stable blood sugar and insulin levels. A 2001 study done in Canada compared the American Heart Association (AHA) low-fat diet with a glycemic-controlled diet to determine the effects of both diets on blood fats and weight.[9]

Volunteers were slightly overweight. At different times, they ate either (1) a glycemic controlled diet containing low glycemic index (GI) carbohydrates balanced with protein. The food quantities were not restricted; they were allowed to eat enough food to be satisfied on this diet, (2) a standard low-fat American Heart Association (AHA) recommended diet with enough food to be satisfied, or (3) an AHA diet with the caloric intake limited to the same number of calories as each volunteer had eaten on the glycemic control diet. All volunteers ate each diet for six days separated by two-week "off diets" periods.

On the sixth day of each type of diet in the study, the researchers measured blood sugar, insulin, triglycerides, cholesterol and weight for all subjects. The results showed that when the volunteers ate the glycemic controlled diet, they had significantly lower blood sugar, insulin, triglyceride and total cholesterol levels than on the other diets and lost an average of 2.4 pounds in six days. When the volunteers ate an AHA diet with enough food to satisfy hunger, they ate about *one-third more* calories than on the glycemic control diet. At the end of the 6-day test period they had the highest blood sugar and insulin levels, a 28% increase in triglycerides, a 10% decrease in HDL ("good") cholesterol, and they gained an average of 0.2 pounds. When the volunteers ate the calorie-restricted AHA diet, their blood sugar and insulin levels were between those they had on the first two diets; there was no significant change in their blood fats over the course of the 6-day test period, and they lost an average of 1.7 pounds.

---

5 Rose, 29.

6 Rose, 29-30.

7 Rose, 173.

8 Rose, 174.

9 http://www.montignac.com/en/principles-of-the-montignac-method-scientifically-validated/

What I find amazing is that being on a glycemic control diet for just six days produced a significant change in blood fats. This means that it does not take long for a low GI diet to improve the HDL/LDL ratio and thus begin the process of improving blood vessel health. See the previous chapter for how to make this dietary change.

In *Save Your Sight!* Drs. Marc and Michael Rose also offer advice about supplements to improve artery health. They advise taking:

(1) **Antioxidants**[10] to neutralize harmful free radicals before they damage arteries. The antioxidant supplements they recommend include **vitamins C and E, beta-carotene, bioflavonoids, selenium, and N-acetyl-cysteine**, which increases the body's production of the antioxidant glutathione.

(2) **B vitamins, especially B-12 and folic acid**, to keep levels of the amino acid homocysteine low.[11] Twenty studies which included a total of 2000 patients showed that homocysteine levels are higher than normal in people with heart and artery disease.

(3) **Magnesium**[12] to help keep blood flowing smoothly through the blood vessels in the eyes and throughout the body. Magnesium helps keep arteries free of calcium deposits by promoting the absorption and metabolism of the calcium in the deposits. Magnesium also opposes the constricting effects of calcium and helps relax and open blood vessels.

The Drs. Rose cite a study[13] in which volunteers with high blood pressure or decreased heart strength were given intravenous magnesium which lowered blood pressure and normalized heart muscle function. Magnesium also causes the level of HDL "good" cholesterol to increase, regulates the heartbeat and normalizes blood pressure.

Other supplements to add for blood vessel health include:

(4) The amino acids **lysine and proline** which make sticky lipoproteins smooth so they can slip off blood vessel walls.[14]

(5) The amino acid **taurine** which supports the health of blood vessels in the eye.[15] The doctor who originally told my mother about IVs stopping the progression of macular degeneration and gave us the IV protocol advised supplementing with adequate zinc, selenium and taurine when Mark tapered off the frequency and discontinued taking IVs. Details of how much, when and how to take these supplements are on the list of supplements for macular degeneration on page 27.

Now that we have discussed internal factors that affect eye health such as nutrition, blood glucose control, and protecting the blood vessels in the eye, we will move on to external factors such as damaging light, toxins, and drugs in the following chapters.

---

10   Rose, 169-170.
11   Rose, 170.
12   Rose, 170-171.
13   Rose, 295.
14   Rose, 176
15   Rose, 75.

50

Protect Your Eyes from Damaging Light

Less than an hour after we first heard the diagnosis of macular degeneration, a cashier at our health food store told me that Mark's eyes should be protected from damaging light. Never underestimate the value of what you hear from ordinary people who care about everyone who crosses their path. They may know more than the experts and they treat you with kindness in a crisis.

That morning the eye doctor had said Mark should eat kale, salmon and egg yolks. Since our sons would both be having dinner with us and I had an entrée ready to go, I went to the health food store to purchase some kale. I asked the cashier if she knew a way to cook kale for someone who didn't like kale. She didn't have helpful cooking advice but suggested looking at the kale chips the health food store carried. She also told me to keep hardboiled eggs in the refrigerator for snacks and that Mark needed to protect his eyes from both ultraviolet (UV) and blue light.

Nine days later, the retina specialist advised us to purchase "a good brand of sunglasses" and named several brands. She said UV light was damaging to eyes but blue light was not a problem.

As I read books and searched the internet, I discovered that all sources were in agreement about the need for protection from UV light but that most older sources did not mention blue light. (*Save Your Sight!*[1] was an exception with advice to seek dark glasses that protect eyes from blue light in 1998). However, more recent sources also include advice about protecting eyes from high-energy blue light.[2]

Here is some information about light: The light that comes from the sun appears white but is made up of various colors of light. When raindrops or a prism separate sunlight into its various colors, we see a rainbow. The red end of the rainbow is the visible light that has the least energy and longest wavelength. The light at the blue end of the rainbow has the most energy and shortest wavelength. Ultraviolet light has an even shorter wavelength and more energy than blue light. It is invisible to the human eye.

---

1 Rose, Marc R., MD and Michael R. Rose, MD. Save *Your Sight! Natural Ways to Prevent and Reverse Macular Degeneration.* (New York, NY: Warner Books, 1998), 230.

2 Abel, Robert, Jr. MD. *The Eye Care Revolution: Prevent and Reverse Common Vision Problems.* (New York, NY: Kensington Publishing Corp., 1999, 2014), 48-49; Samuel, Michael A., MD. *Macular Degeneration: A Complete Guide for Patients and Their Families.* B(Laguna Beach, CA: Basic Health Publications, 2008), 28-29; Mogk, Lylas G. MD and Marja Mogk. *Macular Degeneration: The Complete Guide to Saving and Maximizing Your Sight.* (New York, NY: Ballantine Publishing Company, 1999, 2003), 102-103; https://www.macular.org/ultra-violet-and-blue-light; http://www.webrn-maculardegeneration.com/blue-blocker-sunglasses.html; http://www.allaboutvision.com/cvs/blue-light.htm.   .

Because ultraviolet light has the greatest energy, it would seem to have the most potential to damage eyes. However, the cornea and lens of the eye absorb most of the UV light that enters the eye. Visible blue light is not absorbed and thus reaches the retina. Therefore, high energy visible blue light is the type of light most likely to cause significant damage to the retina and especially to the macula.[3]

The American Macular Degeneration Foundation reports that the people most at risk of macular degeneration caused by light are those who spent five or more hours outside most days in their teens, twenties, and thirties.[4]

The current belief about how light damages the retina is that the process is mediated by a waste product called lipofuscin which is shed from the rods and cones as they slough off dead cells. When hit by light, lipofuscin produces damaging free radicals.[5] Excellent eye nutrition provides antioxidants that can neutralize free radicals. Additionally, the pigments lutein and zeaxanthin plus carotenoids are protective for the retina and macula.[6]

Dr. Giuseppe Querques, head of the Ophthalmology Department at the University of Milan, has directed studies on the effect of blue light on the macula and found that blue-violet light with a wavelength between 415 and 455 nanometers is particularly harmful. He recommends protective dietary supplements such as those listed in the last paragraph and shielding the eyes from the blue-violet rays of the sun with sunglasses that block high-energy blue light as well as ultraviolet light.[7]

In *Save Your Sight!* ophthalmologists Marc and Michael Rose, MD, point out that some sunlight is needed for good health. Vitamin D is made in the skin when it is exposed to UV light. Additionally, exposure to the light and dark cycles of day and night controls the production of melatonin which, when secreted in a light-responsive cycle, helps to promote good-quality sleep.[8] However, the authors recommend moderation in exposure to sunlight and wearing high-quality sunglasses if you will be outside for more than 15 minutes, especially between 10 am and 2 pm.[9] Their advice for purchasing sunglasses is to get lenses that block 100% of UVA and UVB rays and at least 85% of blue-violet light. The Oakley™ sunglasses we purchased (one of the brands recommended by the retina specialist) have lenses made of a substance called plutonite, which is a special polycarbonate that blocks 100% of all types of ultraviolet light plus 100% of blue-violet light. Mark says he does not see

---

3 Samuel, 26-27.

4 "Ultraviolet and Blue Light Aggravate Macular Degeneration." https://www.macular.org/ultra-violet-and-blue-light

5 Rose, 57 and http://www.pointsdevue.com/video/blue-light-interview-prof-giuseppe-querques-italy-highlights

6 Rose, 57-58.

7 I rarely watch online videos, but this video http://www.pointsdevue.com/video/blue-light-interview-prof-giuseppe-querques-italy-highlights of an Italian ophthalmologist has English subtitles so is both informative and enjoyable to listen to.

8 Rose, 22.

9 Rose, 23.

the color blue as much with these glasses on and that glare is eliminated, making his eyes more comfortable as well as eliminating squinting.

Beware of fashion sunglasses. They can do more harm than good. Their plastic lenses block only some of the UV light but the colored tint makes eyes comfortable so there is no squinting and the pupils open widely. Therefore, more harmful blue-violet light may reach the retina and macula than if no dark glasses are worn.

With optimally protective sunglasses and good eye nutrition, you can still enjoy a picnic, bike ride, walk, outdoor exercise or other outdoor activities. Have fun!

Avoid Eye-Damaging Drugs

Many pharmaceutical drugs directly or indirectly cause macular degeneration to progress much more rapidly than it would if the drug were not taken. For this reason, macular degeneration patients should strive to be drug-free or as close to drug-free as possible. Dietary and lifestyle changes and natural remedies such as vitamins, minerals, other nutrients, and herbs may be explored as options for treating many of the conditions for which eye-damaging drugs are usually prescribed. If a doctor prescribes a new drug, check with a trusted pharmacist to be sure that it is not one of the hundreds of drugs that can make macular degeneration progress rapidly. If the prescribed drug is eye-damaging, ask the doctor or pharmacist for other less-damaging options which will treat the problem while you work toward eliminating drugs and replacing them with natural remedies. **The list of problematic drugs in this chapter[1] is not exhaustive, so if you are currently taking any drugs, whether prescription or over-the-counter drugs (no-prescription access does not mean they are safe), check with the pharmacist about them.**

There are many prescription and over-the-counter drugs which cause damage to the eyes, some directly and some by making the fragile retina and macula more easily damaged by light. There are also drugs that adversely affect nutritional status. These are harmful because ideal nutrition for eyes is the best way to prevent macular degeneration from progressing.

The remainder of this chapter lists types of drugs to avoid and why to avoid them. Although the main concern here is the effect these drugs have on macular degeneration, other types of side effects are mentioned for two reasons: (1) you may wish to be warned about all side effects or (2) you, or those you care about, may have experienced side effects without realizing they were caused by the drugs.

Cholesterol-lowering drugs (statins) are some of the most commonly prescribed drugs, and although they may improve numbers on blood tests, they do not achieve their ostensible goal of lowering the incidence of heart attacks and strokes.[2] Instead, they worsen health by causing calcification in arteries which may lead to more serious heart and artery problems. Increased calcification occurs all over the body, including in the delicate arteries of the retina and macula.

While statins do decrease the levels of cholesterol on a patient's blood test, this does not mean the patient's heart health is better. Rather, it means that there is an increased risk of dying from a heart attack. In a study published in *Atherosclerosis*, 6,673 users of statins who had no previously known coronary artery disease (i.e. they were taking statins preventatively) had coronary CAT scan angiography

---

1 Rose, Marc R., MD and Michael R. Rose, MD. Save *Your Sight! Natural Ways to Prevent and Reverse Macular Degeneration.* (New York, NY: Warner Books, 1998), 181-193.

2 Mercola, Joseph, DO. The Cholesterol Myths That May Be Harming Your Health. October 21, 2011. http://mercola.ebeaver.org/2011/10/22/the_cholesterol_myths_that_may_be_harming_your_health/

(CCTA) which enabled the researchers to see their coronary arteries and determine the composition of plaque in the arteries. **Patients taking statins had a 52% increase in presence and extent of calcified coronary plaque compared to those not taking statins.**[3]

A study in *Diabetes Care* showed that diabetics with advanced artery disease and taking statins had significantly more calcification in their arteries than non-statin taking diabetics. In those who began taking statins during the course of the study, progression of coronary and abdominal aorta calcification increased significantly when they began using statins. Calcification is dangerous in major arteries; they become stiff and inflexible when lined with calcium deposits, and individuals are **more likely to experience a coronary artery blockage (heart attack) or abdominal aortic aneurysm, which is usually fatal.**[4] (The exception to "fatal" is if this condition is discovered when asymptomatic and sugically corrected).

Statins also have a myriad of other side effects, many of which are serious. Macular degeneration patients should avoid statins because **fat soluble vitamins and other nutrients such as the retina-essential pigments lutein and zeaxanthin, carotenoids, bioflavonoids, and other nutrients that can save your sight will be poorly absorbed, transported to the eyes, and utilized if the patient takes statin drugs.** Rather than taking these drugs, see pages 48 to 49 for how to use natural and nutritional methods to improve HDL/LDL cholesterol ratio, which is the most significant blood test number for artery health. **Other common side effects of statins include nerve, liver, or kidney damage, dementia and increased incidence of diabetes, cancer and depression.**[5] For more about statins and a list of additional side effects, visit this webpage: http://www.healingbasics.life/statins.html

Blood-pressure lowering drugs also are widely prescribed and **interfere with the good nutritional support which your eyes need.** Diuretics cause the depletion of many minerals. Some diuretics cause the retention of calcium, which causes blood vessels to constrict.[6] If you have macular degeneration, you should avoid drugs which cause the blood vessels in your eyes to constrict to a smaller diameter than their already small size.

Drugs that interfere with the absorption of nutrients include antacids (like Tums™ and Mylanta™) and proton pump inhibitor drugs (PPIs) such as Nesxium™, Prilosec™, Prevacid™, and similar drugs. They all cause incomplete digestion and absorption of foods, which adversely affects the

3 Nakazato, Ryo, Gransar, H., et al. "Statins use and coronary artery plaque composition: results from the International Multicenter CONFIRM Registry." *Atherosclerosis.* 2012 Nov;225(1):148-53. doi: 10.1016/j.atherosclerosis.2012.08.002. Epub 2012 Aug 24.

4 Saremi, Aramesh, Bahn, G, et al "Progression of vascular calcification is increased with statin use in the Veterans Affairs Diabetes Trial (VADT)." *Diabetes Care.* 2012 Nov;35(11):2390-2. doi: 10.2337/dc12-0464. Epub 2012 Aug 8.

5 Rose, 184 and Roberts, Barbara H., MD. *The Truth About Statins.* (New York, NY, Pocket Books, Simon & Schuster, 2012), 46-71.

6 Rose, 182.

absorption of all nutrients and will most dramatically **reduce the absorption of minerals. Trace minerals and zinc are especially important for macular degeneration patients, and PPIs decrease their absorption.**[7]

PPIs cause other serious side effects by suppressing the secretion of stomach acid. Since minerals are especially difficult to absorb without sufficient stomach acid, **PPI users are more likely to have bone fractures due to reduced absorption of calcium and other minerals**. There is a higher incidence of **pneumonia** and other infectious diseases among those taking PPIs. This is because stomach acid is the first line of defense against bacteria entering our bodies by way of the digestive system. These drugs also lead to increased risk of **heart problems, kidney disease, and dementia.**[8]

Two of the more serious side effects of PPIs have been reported in the *Journal of the American Medical Association*. A data analysis study was done on elderly people who were initially free of dementia. After seven years, **those who had taken PPIs were 52% more likely to have developed dementia** than those who did not take PPIs.[9] **Chronic kidney disease also increased 20 to 50% with PPI use**. The incidence was higher in those who took PPIs twice daily than those who took them once daily.[10]

PPIs are frequently prescribed for gastro-esophageal reflux disorder (GERD) which occurs when the irritating contents of the stomach enter the esophagus. Stomach acid is necessary to signal the pyloric valve between the stomach and the esophagus to close. What is needed for GERD is not acid-suppressing drugs, but a hydrochloric acid supplement taken with every meal and snack so the pyloric valve closes well.

Heartburn commonly occurs as we age because of the decline in our ability to make enough hydrochloric acid (HCl) after a large meal when it is most needed. Since the signal to "make acid" persists for hours, we end up with too much stomach acid several hours later, often in the middle of the night. Natural remedies such as bitters or apple cider vinegar before a meal and dietary changes may be used to help heartburn.[11] Although it seems counter-intuitive, the most effective natural treatment for heartburn is to take a hydrochloric acid supplement with meals to cause the pyloric valve to close, facilitate digestion, and turn off the "make acid" signal that causes secretion of HCl much later, thus producing heartburn. Consult your holistic medical practitioner about whether and how you should take hydrochloric acid supplements. Other supplements such as slippery elm can increase comfort

---

7 Rose, 84.

8 Kresser, Kris. Proton Pump Inhibitors: So Dangerous That Prescriptions Border on Being Criminal. June 14, 2016. https://www.sott.net/article/320501-Proton-Pump-Inhibitors-So-dangerous-that-prescriptions-border-on-being-criminal

9 Gomm, W., Von Holt, K., et al. "Association of Proton Pump Inhibitors With Risk of Dementia: A Pharmacoepidemiological Claims Data Analysis." JAMA Neurology, 2016 Apr;73(4):410-6. doi:10.1001/jamaneurol.2015.4791.

10 Lazarus, B, Chen, Y, et. al. "Proton Pump Inhibitor Use and the Risk of Chronic Kidney Disease." JAMA Internal Medicine. 2016 Feb;176(2):238-46. doi: 10.1001/jamainternmed.2015.7193.

11 Rose, 136-142.

while in the process of adding HCl as you begin to take it or if you forget to take sufficient HCl. Slippery elm is most effective when taken as the powder mixed into water so it can coat the esophagus.

Non-steroidal anti-inflammatory drugs (NSAIDS) such as aspirin, ibuprofen, (Motrin™, Advil™), celecoxib (Celebrex™), naproxen (Aleve™), prescription arthritis drugs and similar drugs can be photosensitizing, meaning that they **make sunlight more able to damage eyes, and they may also cause bleeding from the choroidal capillaries which underlie the macula,** which accelerates wet macular degeneration. Therefore, **all NSAIDS, both prescription and over-the-counter, should be avoided by macular degeneration patients.**[12]

What can we do for pain instead of using drugs? The most effective treatment is to treat the problem causing pain. Dr. Leo Galland reports that having patients avoid all foods to which they are allergic can eliminate arthritis pain and the need for drugs.[13] For other types of joint and muscle pain, seek treatment from a physical therapist who pays attention to the whole body, not just the painful area. Sometimes postural problems or issues with neighboring joints can affect the painful area even if the original problem was due to an injury. Then diligently do the exercises the therapist prescribes.

There are many natural remedies for pain; this list is not exhaustive. For sore muscles or injuries, application of cold (initially) or heat can help. A half-hour soak in a bathtub of warm water with one cup of Epsom salts relaxes achy muscles and soothes joints. Herbal remedies taken regularly, such as feverfew for migraine headaches, can reduce dependence on drugs. Acupuncture is also effective for pain.

I have found homeopathic arnica effective for pain. Before my mastectomy, I told a nutritionist about my desire to avoid drugs. She told me that when she had surgery, she had her family slip her arnica beginning as soon as possible after surgery. She said arnica could be used as often as every half hour and then less frequently when the pain does not return in a half hour. She also told me what drug to ask for instead of the more "potent" narcotics if needed. (It wasn't needed). I did very well with the arnica, an intravenous form of ibuprofen which was started at the conclusion of the surgery and continued through my twenty-two hour hospital stay, and Tylenol™ for about two weeks at home. Since then I have used arnica for muscle and joint pain and arthritis.

Homeopathic remedies come in different strengths depending on how they are diluted; which strength is needed may vary between individuals and with the condition treated. I tried homeopathic arnica for surgical-type pain when I had a 2½ hour needle biopsy to remove all calcifications from the non-affected breast to determine if it also needed to be removed. The 200C strength was effective then so I took it for my surgery.

---

12 Rose, 187-189.

13 Interview with Leo Galland, MD by Marjorie H. Jones, RN., "Leaky Gut – What Is It? What Factors Cause It? What Can Be Done?" *Mastering Food Allergies Newsletter,* #86, July/August 1995, 2.

For arthritis and various for aches and pains I use the 30C strength of arnica. Homeopathic combinations for pain or arthritis also are helpful.

Finally, consume anti-inflammatory foods and supplements regularly. Omega-3 fatty acids and boswellia taken daily can reduce inflammation and chronic pain. Also see the list of anti-inflammatory foods on pages 196 to 197 and the "Ginger Tea" and "Gingerale" recipes on pages 89 and 90.

Anticoagulant drugs (Coumadin™, heparin, etc.) can lead to bleeding from the capillaries that nourish the macula. Some anticoagulants are also photosensitizing drugs, meaning that sunlight will be much more damaging to your eyes if these drugs are taken.[14]

Hormone replacement drugs such as **synthetic estrogens** (Premarin™), **androgens** and **corticosteroids** (Prednisone™) can **cause blood clotting which can interfere with blood flow in the eyes.**[15]

A large number of other drugs are photosensitizing. This list is not exhaustive, so **ask your doctor and/or pharmacist about all the drugs you take.** A partial list of photosensitizing drugs includes[16]:

Diuretics of several kinds
Glaucoma drugs
Drugs used to treat irregularities in the heartbeat
Antihistamines
Antibiotics and sulfa drugs
Anti-anxiety and anti-depression drugs
Sulfonylurea diabetes drugs
Drugs which treat irritable bowel syndrome
Retin-A drugs for treating acne, fine wrinkles and skin discoloration
Coal-tar drugs for treating dandruff, psoriasis and seborrheic dermatitis
Oral contraceptives

Some drugs can directly damage the retina. They include **NSAIDS** which are listed on the previous page as well as[17]:

Plaquenil™ for rheumatoid arthritis
Catapres™ for high blood pressure
Terbenafine™, an anti-fungal drug
Thoridazine™, an anti-psychotic drug
Mefloquine™, an anti-malarial drug

and others drugs. **Ask a trusted pharmacist about every drug you take.**

---

14 Rose, 186-187.
15 Rose, 188-189.
16 Rose, 184-186.
17 Rose,187-188,

Drugs are toxins that are removed from your body by detoxification in your liver and other means. The herb milk thistle supports the liver in the detoxification process so is helpful taken with drugs. For more about how to remove drugs and other toxins from your body, see the next chapter.

Toxins : How to Avoid Them and Remove Them from Your Body

Most of us enjoy comfort in our lifestyles and living environments. We walk on plush carpet in our homes and travel wherever we want quickly by automobile or airplane. This lifestyle costs us in non-monetary ways. The synthetic materials that make up our homes out-gas formaldehyde and a myriad of additional toxic chemicals. Herbicides, pesticides and chemical fertilizers keep our lawns looking beautiful and our food profitably produced, or so their manufacturers say, but they are harmful to our bodies. Our cars, factories and airplanes pollute the air we breathe. These exposures are toxic to our bodies and especially to fragile eyes.

Although environmental toxins can be difficult to avoid, it is potentially possible to control exposure to toxins in our food. Since the stakes are high for macular degeneration patients, if finances permit, you may want to consider trying to eliminate some of the toxic substances you are ingesting. Although a diet high in naturally raised meat and poultry can be pricey, by eating organic produce we can affordably avoid pesticides, herbicides and plant GMOs. Hormones in dairy products can be avoided by choosing reasonably priced store-brand milk that is labeled with something like, "Our farmers pledge not to treat their cows with rBGH or rBST." In our area, Kroger, Safeway and Target stores all have this statement on their store brand milk. Although cottage cheese and dairy products other than milk do not have the same statement on their labels, the store brand of these products is likely to come from the same cows as the store-brand milk. Therefore, it seems likely that Kroger, Safeway and Target store brand cottage cheese, cheese, etc. may be safer than other brands of dairy products. If you want to be completely sure that you are getting pure milk and dairy products, look for brands labeled "organic."

If we read food labels we can avoid some toxic ingredients added to our foods such as monosodium glutamate (MSG) and artificial sweeteners such as aspartame. MSG and aspartame are excitotoxins which over-stimulate brain cells and cause the death of neurons. MSG also stimulates taste receptors on the tongue so foods seem to taste much better and therefore is added to almost all processed foods to increase appetite for the food and profit. MSG is disguised on food ingredient lists as hydrolyzed vegetable protein, hydrolyzed yeast, natural flavorings, spices, and many other names.[1]

In *Excitotoxins: The Taste That Kills,* Russell Blaylock, MD, tells of a study in which infant mice were fed MSG to learn more about hereditary retinal dystro-

---

1 Rose, Marc R., MD and Michael R. Rose, MD. Save *Your Sight! Natural Ways to Prevent and Reverse Macular Degeneration.* (New York, NY: Warner Books, 1998), 208-209 and Lant, Carla. "Excitotoxins: The FDA-approved Way to Damage Your Brain." June 22, 2015, Honey Colony. https://www.honeycolony.com/article/excitotoxins-fda-approved-damage-brain/ . Also see page 201 for additionaly names MSG is disguised as on ingredient lists.

phy. When the ophthalmologists sacrificed the mice and examined their retinas microscopically, they saw something they did not expect. All of the neurons in the inner layers of the retinas had been destroyed. Ten years later the experiment was repeated by John W. Olney, MD, with the same retinal destruction. Additionally, he found that cells in the hypothalamus were destroyed after just one feeding of MSG.[2] Macular degeneration patients are advised to avoid processed foods because they are nearly devoid of critical nutrients, but they should also be avoided because they are likely to contain MSG in some form. See page 201 for a long list of the names MSG may be hiding under on ingredient lists so you can avoid these disguised forms of MSG in foods.

In *Save Your Sight!* ophthalmologists Marc and Michael Rose, MD advise macular degeneration patients to **avoid aspartame (NutraSweet™) which is a retinal and optic nerve toxin**.[3] In *The Eye Care Revolution*, Dr. Robert Abel prescribes **avoidance of all artificial sweeteners**[4]. (There were several on the market when he revised this book in 2014. In 1998 when *Save Your Sight!* was written, aspartame was the only widely-used artificial sweetener). He says all artificial sweeteners are toxic to nerves in the eye, to nerves throughout the body and to the brain. He reports that diabetics who use them are much more likely to suffer from neuropathy than those who do not.[5] He also says that his patients who eliminate diet sodas say that their eyes seem much better.[6]

Dr. Abel also gives advice about inhaled toxins, such as not to jog along a busy highway, thus breathing exhaust from the vehicles, and to avoid second hand smoke.[7]

Although Drs. Marc and Michael Rose say, "Obviously, it isn't practical for you to try to avoid all potential toxins,"[8] there are ways to decrease the load of toxins your body carries and lessen your exposure to toxic substances. Unless your home is brand new, hopefully most of the building materials are well out-gassed. Air out your home as often as is convenient in order to dispel toxic fumes that you may not even smell from plastics, black-out drapery liners, furnishings and cleaning products.

As furnishings and other items in your home wear out, replace them with less toxic alternatives. For example, replace carpet with hardwood that is pre-finished with polyurethane and which should never need refinishing. Choose new furniture that is not made with formaldehyde-containing particle board or plywood or other toxic materials. We recently purchased a sofa from DutchCrafters™ made by Amish craftsmen using all natural hardwood and are very pleased with it. An option we've

---

2 Blaylock, Russell, MD. *Excitotoxins: The Taste That Kills. (Santa Fe, NM, Health Press, 1995).* xviii-ix, 215.

3 Rose, 208-209.

4 Abel, Robert, Jr. MD. *The Eye Care Revolution: Prevent and Reverse Common Vision Problems.* (New York, NY: Kensington Publishing Corp., 1999, 2014), 279.

5 Abel, 211.

6 Abel, 255.

7 Abel, 332.

8 Rose, 207.

used many times in the past was to purchase unfinished furniture and finish it with AFM Safecoat™ stain and polyuraseal. See "Sources," page 220 for more information about these products and the Amish furniture.

When you buy an item of new clothing, remove the chemicals it is laden with before wearing it. I have always washed new clothing before wearing it, but recently read about a new twist on this. In *The Allergy Solution*, Dr. Leo Galland recommends soaking new clothing overnight in the washer with warm water first, and the next day washing it repeatedly until it smells all right.[9] I was surprised when I soaked a rust-colored turtleneck top overnight and the water was deep orange the next morning.

For best health, use fragrance-free laundry products. What is wrong with fragrance if you are not sensitive to it? It is there to cover the smell of toxic cleaning ingredients. Furthermore, chemical fragrances are themselves toxic. A study was done on what was in the air that clothes dryers vented to the outside. Laundry done with fragranced detergents produced twenty one harmful volatile organic compounds (VOCs). If both fragranced detergents and fabric softening dryer sheets were used, the list was much longer. There were no VOCs emitted from the laundry washed with unscented detergent only and dried without dryer sheets.[10]

Air fresheners don't actually freshen the air, they just cover unpleasant odors with chemical perfumes that are as toxic as those in laundry products. The use of air fresheners is linked to allergies and worsening asthma.[11] Remove whatever is causing offensive odors, such as garbage in your kitchen wastebasket. Then open the windows to really freshen the air.

Alternatives to chemical cleaning products can be found at health food stores, but read the labels even there. This book does not give recommendations for specific products found at the health food store because, like the packaged foods there, the ingredients change often. Old standbys for non-toxic cleaning include vinegar, baking soda, borax and BonAmi™ scouring powder. When you need serious cleaning power, try some AFM SuperClean™. A bottle of this concentrate will last a long time. The label directs how to dilute it from between 1:15 to 1:2 to clean everything from kitchen counters to walls to difficult-to-remove soap scum deposits. For soap scum, you may want to moisten the surface with 1:2 SuperClean™, let it stand an hour to overnight and then scrub the surface with BonAmi™. For information on where to get SuperClean™, see "Sources," page 220.

Microfiber cloths can eliminate the need to use many cleaning products. For cloths that pick up and hold dust, look for 80% polyester and 20% polyamide fibers. Before purchasing such cloths, feel them. Microfiber that works well contains polyamide and feels sticky. The surface of most microfiber cloths is covered with loops. My current favorite looped cloths which contain 20% polyamide are made

---

9 Galland, Leo, MD. *The Allergy Solution*. (Carlsbad, CA, Hay House, Inc., 2016), 91-92.

10 Galland, 88.

11 Galland, 93.

by VibraWipe™. See "Sources," page 220, for information about purchasing these cloths.

Use water with microfiber cloths when cleaning, either by wetting the cloths or with a spray bottle. Although microfiber can be used dry for dusting, I slightly dampen cloths before dusting to minimize the amount of dust that escapes the cloth and is inhaled. I also change cloths frequently while cleaning and deposit the soiled cloths in the clothes washer as I work. When I finish cleaning, I wash the cloths in cool water with laundry detergent only and then hang them to dry. Do not wash microfiber cleaning cloths with cotton fabrics or they will pick up permanent lint thus destroying their effectiveness for picking up dust.

Personal care products should be natural and unscented, not petroleum-based and should not contain parabens or phthalates.[12] Shop for them at health food stores and read the labels. Interestingly, recent studies have linked the rapidly rising incidence of peanut anaphylaxis in children to early exposure to peanut oil in skin care products.[13] Thus, be careful of what you put on your skin because it will be absorbed. The practice of slathering people, especially children, with toxic DEET to prevent mosquito bites is ludicrous considering that 100 mg of vitamin B1 taken before exposure works better and is non-toxic. For more about using vitamin B1 to prevent mosquito bites, see this webpage: http://www.healingbasics.life/mosquito-repellents.html

Although it can be nearly impossible to control toxins are in the outdoor air, in our workplaces and in stores, we do have some control of what is in our homes. Awareness of what is toxic and of alternative products and non-toxic methods we can use instead will lower our personal load of toxins and help spare our eyes damage from toxins.

What about the toxins that our bodies are already carrying? Drs. Marc and Michael Rose, MD, recommend occasional fasting, sweating in a sauna, chelation IVs or herbal cleansing regimens.[14] Since our family has little experience with most of these methods I will refer you to *Save Your Sight!* for more information on them. The Drs. Rose recommend using UltraClear™[15] instead of using water only for fasting longer than a few days. My allergy doctor once recommended that I try Ultra-Clear™. Since it is made mostly from rice with many additional nutrients and I am allergic to rice, he gave me a rice neutralizing solution to use to prevent reactions. It did not work, possibly because of allergenic sources of the added nutrients. Whatever the reason, those with food allergies and gluten intolerance need to exercise care with what products they use for fasting. I use a water fast of about 68 hours duration each time I take an LDA treatment (mentioned on page 12), eat sparingly before and after the fast, and find that it weakens me greatly. However, ask your

---

12 Servan-Schreiber, David, MD, PhD. *AntiCancer: A New Way of Life*. (New York: Penguin Group, Inc., 2009), 6 of "AntiCancer Action" section in the center of the book.

13 Galland, 49.

14 Rose, 206-207, 220-225.

15 Rose, 221,

doctor about fasting and other ways of cleansing. If you are able to use any of the cleansing techniques in *Save Your Sight!* you may find them helpful. The Drs. Rose especially recommend IVs for chelation and report that their macular degeneration patients who take chelation treatment for heart and artery disease find that their vision also improves.[16]

Mark's naturopath told him to use a cleansing technique that most patients can use safely and inexpensively, which is to drink sufficient water. Her recommendation was to drink half his body weight in ounces of water per day. In *The Seven Healers,* Dr. Scott Conrad, MD, says that this formula for how much water is needed is a good place to start, but for some people it is still not enough. He also says that thirst can be an unreliable indicator of how much water we need, especially as we age.[17] His method for individually determining how much water is enough is to observe the color of your urine. If it is very pale to clear, you are drinking enough water.[18] He says to aim for "pitch clear" but notes that if you take B vitamins, which make urine fluorescent yellow, this may interfere with using urine color to determine how much water to drink.

Dr. Conrad explains why it is so important to be well-hydrated and how sufficient water removes toxins from our body. He says each of our body's cells is like a house on a canal with water running on all sides of it. The canals between and around our cells are filled with water. (The water carrying dissolved substances between the cells is called interstitial fluid). When the cells' metabolism creates waste products or the environment outside our bodies introduces toxins, these substances are dumped into the canals to be carried off. When the fluid moves well, with plenty of water coming in and flushing the canals out, the removal of waste products from cellular metabolism and toxins from external sources happens efficiently and quickly. If water supply is insufficient and the canals are not flowing freely, waste products and toxins can build up and be recirculated rather than being washed into the lymphatic system and blood vessels to be carried off to the liver and kidneys for detoxification and elimination. When toxins are recirculated rather than eliminated, they can build up and affect the eyes.

It makes sense that the water we drink in large amounts in order to flush potentially harmful toxins out of our bodies should not contain additional toxins that will be added to our bodies. The Drs. Rose reported that much of America's water supply was polluted when they wrote *Save Your Sight!* in the 1990s, and with the use of many new agricultural chemicals, climate issues, etc. the quality of our water has continued to decline.[19] They cited pollution being common with aluminum, lead, chlorine (usually added in the water purification process), fluoride, petroleum pollutants, and parasites such as *Giardia* and *Cryptosoridium*. (Both of these parasites can survive the water purification and disinfection processes used by water treat-

16 Rose, 214-220.

17 Conrad, Scott, MD. *The Seven Healers.* (Dallas, TX: Rapha7ven, 2011), 30

18 Conrad, 32.

19 https://www.cnbc.com/2016/03/24/americas-water-crisis-goes-beyond-flint-michigan.html

ment plants). Purified water bottled in plastic bottles contains chemicals from the plastic that are toxic and may contribute to cancer.[20] They recommend that their patients with eye disease invest in a home water filter.[21]

After I learned about the importance of drinking water, I joined Mark in drinking half my body weight in ounces of water per day. About two hours into my first day of drinking more water, I became insatiably thirsty. My mouth was dry no matter how much water I drank. I think that once my body got the message that there was plenty of water being supplied to it, it wanted more, enough to cleanse out toxins as rapidly as it could. Although I drink my calculated amount every day, mostly in the mid-morning and mid-afternoon, there are still times when I am thirsty for more. In that case, I listen to my thirst and drink more, although I try to drink only enough to satisfy thirst near bedtime.

Do not decrease water intake because it necessitates more bathroom trips. A friend told me that this would be a temporary inconvenience, and she was correct. After about two or three weeks of drinking more water, both Mark and I found that our bladder capacity had increased.

As with moderate exercise, which will be discussed in the next chapter, drinking plenty of cool, pure-tasting filtered water will grow on you until you may want to drink nothing else. Drink up to your health!

---

20 Servan-Schreiber, David, MD, PhD. *AntiCancer: A New Way of Life.* (New York: Penguin Group, Inc., 2009), 86.
21 Rose, 8-9.

Exercise

Circulation in the blood vessels of the eyes is often impaired with macular degeneration. One of the best and safest ways to improve circulation to all parts of your body, including your eyes, is to do moderate exercise. You do not need to join a gym or buy special clothes or equipment. Any kind of exercise you enjoy will improve your health. Walking is free and helps as much as other types of exercise. Like drinking water, it will grow on you, especially if you walk in a park or natural area, and you will enjoy walks.

Regular moderate exercise has many benefits.[1] The most important benefit for macular degeneration patients is **improved circulation to the eyes**. Both your heart and lungs will perform more efficiently when you exercise regularly, which means better waste product removal, nutrient delivery, and oxygenation of the delicate tissues of your eyes.

A second benefit is that **exercise increases HDL "good" cholesterol which results in less potential obstructions in blood vessels including the choroidal capillaries of the eyes** which nourish the macula.[2]

Diabetics, pre-diabetics and individuals with insulin resistance also derive benefit from exercise. Their **blood sugar is better controlled when they exercise regularly, which protects the eyes from harmful blood sugar surges**. Dr. Rob Thompson, a cardiologist who himself developed diabetes, found that if he goes "a couple of days without at least walking thirty minutes, the effects on my blood sugar are immediate and substantial, enough to make the difference between good and poor control of my diabetes."[3]

Finally, **high blood pressure can be improved with regular exercise.** This protects the delicate blood vessels in the eye from the detrimental effects of high blood pressure as well as helping you avoid eye-damaging blood pressure medications.[4]

Many people embark on an exercise program because they want to lose weight. In *The Glycemic Load Diabetes Solution*, Dr. Rob Thompson reports that daily walking or other exercise that uses the leg muscles (walking, swimming, bicycling, etc.) will increase the sensitivity of the muscles to insulin and thus **diminish insulin resistance for one and a half to two days after exercising[5] which can help weight loss**. He made one lifestyle change – he began walking to work – and

---

1 Rose, Marc R., MD and Michael R. Rose, MD. Save *Your Sight! Natural Ways to Prevent and Reverse Macular Degeneration.* (New York, NY: Warner Books, 1998), 202.

2 Rose, 202.

3 Thompson, Rob, MD and Dana Carpender. *The Glycemic Load Diabetes Solution.* (New York, NY: McGraw-Hill, 2012), 195.

4 Rose, 202.

5 Thompson and Carpender, 138-139.

then lost weight without making any other changes.[6] In addition to being enjoyable, walking helps normalize weight.

One caveat about exercise is not to get so concerned about weight that you **over-exercise**, which **causes the release of adrenal hormones and may cause depressed immunity.** On the other hand, **moderate exercise enhances immunity.**[7] Aerobic exercise done for longer than twenty-five minutes or **overly strenuous exercise, especially without food, can stall weight loss rather than helping it.**[8] In *Save Your Sight!* ophthalmologists Marc and Michael Rose, MD, also report that **frequent intense exercise can cause the production of free radicals which can damage the retina of the eye.**[9]

Once you develop an enjoyable exercise routine, you may notice that your mental state improves as well as your physical health and will not want to miss your exercise time. Give it a try, and give your eyes what they need.

---

6 Thompson and Carpender, 141-142.

7 Rose, 6.

8 Hart, Cheryle R., MD and Mary Kay Grossman, RD. *The Insulin Resistance Diet.* (New York, NY: McGraw-Hill, 2008), 192.

9 Rose, 6.

Peace

At our first visit with the naturopath Mark saw at the clinic where he received IV therapy for macular degeneration, she told him that he needed to work on making his parasympathetic nervous system (which promotes "rest and digest") active most of the time and try to tone down the "fight or flight" sympathetic nervous system. He said, "I need to get less compulsive about work," and he has made progress on doing this most of the time.

Although changing one's mindset, as he did, is the best over-arching way to increase active time of the parasympathetic nervous system, there are a number of activities and habits that are also helpful which will be discussed in this chapter.

In *Save Your Sight!* ophthalmologists Marc and Michael Rose, MD, have two recommendations along these lines. The first is to cultivate awareness. Notice what seems to help your health. They recommend developing an awareness of how what you eat and activities such as exercise, talking to a friend and watching TV make you feel.[1] In the twenty-first century, we need to be especially aware of how excessive "screen time" makes us feel compared to an activity such as reading a good book or walking through a park or in a natural area.

As this awareness leads to engaging in activities that help your health and eyes, it also empowers you to regain control over your health and will reduce feelings of helplessness. Moving away from "helpless" improves medical outcomes as shown by numerous studies.[2]

Surprising the researcher, a study demonstrated this for women with metastatic breast cancer. There were two groups of women; one group did not attend support groups but the women in the other group attended weekly support group meetings of eight to ten women. About ten years after they'd been diagnosed with cancer and participated in the study's support groups, the researcher conducted a follow-up survey. The researcher was amazed that three of the fifty women who attended the support groups answered the phone themselves when he called. No women in the control group had survived that long. The support group women who did not survive lived twice as long as the control group women. There was even a difference between those who went to the support group regularly versus sporadically. The more often a woman attended, the longer she lived. This study showed that doing something to help oneself, such as attending a support group, increased survival time.[3] To read more about similar studies, visit this webpage: http://healingbasics.life/why-you-must-help-yourself.html#SCR

---

1 Rose, Marc R., MD and Michael R. Rose, MD. Save *Your Sight! Natural Ways to Prevent and Reverse Macular Degeneration.* (New York, NY: Warner Books, 1998), 4.

2 Servan-Schreiber, David, MD, PhD. *AntiCancer: A New Way of Life.* (New York: Penguin Group, Inc.,2009), 159.

3 Servan-Schreiber, 156. Also Spiegel, D, et al., "Effect of Psychosocial Treatment of Survival of Patients with Metastatic Breast Cancer." *Lancet.* 2:8673. Nov. 18, 1989: 1209-1210.

In *Save Your Sight!* Drs. Marc and Michael Rose, MD, also recommend, "Have a spiritual practice that gives you a sense of a higher meaning to life. When you have a sense of purpose, of meaning in your life, everything you do is enhanced by that point of view. All the better if your spiritual practice includes some form of meditation."[4] A positive outlook and mindset can make life much more pleasant and its support of parasympathetic nervous system activity enhances your body's ability to heal.

Numerous techniques for meditation exist, including prayer. Many of the techniques involve breathing. Cancer patients have better outcomes when they practice meditative breathing as taught by Dr. David Servan-Schreiber and used in his own cancer recovery. In his book *AntiCancer,* he says that you do not have to believe in anything to profit from meditative breathing's health benefits and relaxing effect.[5]

Here is Dr. Servan-Schreiber's beautifully written description of how to practice meditative breathing: "Begin by sitting comfortably, in what the Tibetan master Sogyal Rinpoches calls a 'dignified' posture. It gives full freedom to the flow of air that slips down through the nostrils toward the throat, then the bronchi, and finally to the bottom of the lungs, before reversing its route. With your attention focused, take two deep, slow breaths to begin relaxation. A sensation of comfort, lightness, and well-being will settle into your chest and shoulders. As you repeat this exercise, you will learn to let your breathing be led by your attention, and to let your attention rest on your breath. As you relax, you may feel your mind become like a leaf floating on water, rising and falling as waves pass underneath. Your attention accompanies the *sensation* of each intake of breath and the long exhalation of air leaving the body gently, slowly, gracefully, all the way to the end, until there is nothing more than a tiny, barely perceptible breath left. Then there is a pause. You learn to sink into this pause, more and more profoundly. It's often while resting briefly in it that you feel in the most intimate contact with your body. With practice, you can feel your heart beating, sustaining life, as it has been doing indefatigably for so many years. And then, at the end of the pause, notice a tiny spark light up all by itself and set off a new cycle of breath. What you feel is the spark of life, which is always in us and which, through this process of attention and relaxation, you may discover for the first time.

"Inevitably, your mind is distracted from this task after a few minutes and is drawn toward the outside world: the concerns of the past or the obligations of the future. The essential art of this 'radical act of love,' consists of doing what you would do for a child who needs undivided attention. You recognize the importance of these other thoughts, but while patiently promising to attend to them when the time comes, you push them to the side and come back to the person who really needs you in the present moment, that is, yourself."[6]

---

4 Rose, 5.

5 Servan-Schreiber,166.

6 Servan-Schreiber,164-165.

By practicing this daily for at least ten minutes, his patients can bring coherence to their biological rhythms. That means their heart rate, blood pressure, respiration, and other functions all cycle in synchronization with each other. This results in better immune function, less inflammation, and better regulation of blood sugar levels.[7]

Another way to relax and activate the parasympathetic nervous system is to spend time in nature. If there is a path through a natural area or a park nearby, take a walk there as often as you can. This is a good way to make daily exercise a health-promoting treat for your mind as well as your body.

Other activities that activate the parasympathetic nervous system include playing with children or pets, massage, and revisiting good memories that elicit positive emotions. Any thoughts or activities that make you feel grateful, loving or contented will usually activate the parasympathetic nervous system.[8]

If your sympathetic nervous system becomes activated by stresses of modern life such as being in a traffic jam that is making you late, a frightening experience, worries, or frustrating circumstances, there are activities that help turn off the "fight or flight" response of the sympathetic nervous system. These include yawning and humming. Humming stimulates the vagus nerve, which is the major nerve of the parasympathetic nervous system. I like to hum hymns because, in addition to the vibration stimulating the vagus nerve, the words are calming. I heard that soldiers are trained to inhale to a count of five seconds and then exhale to a count of fifteen seconds in dangerous or frightening situations. After these techniques have begun to relax you, you may be able to think through the situation and reach a positive mindset to further activate the parasympathetic nervous system.

Although the techniques and activities above are helpful for reducing sympathetic nervous system activity, what you do with your mind long-term is what really counts to help you achieve peace. Skip to pages 174 to 176 for more about a deeper level of peace, or read the next chapter about Mark's progress and peruse the recipes that follow on your way to the rest of the story about peace.

---

7 Servan-Schreiber,168.

8 Zmijewski, Chrissy. "Activate the Parasympathetic Nervous System to Improve Recovery." PTonthenet, November 26, 2014. http://www.ptonthenet.com/articles/activate-the-parasympathetic-nervous-system-to-improve-recovery-3910

Progress

Progress is a difficult concept for macular degeneration patients. They are told that the best they can expect is for the disease to advance slowly and thus their eyes might serve them for activities such as dressing and walking for some time. Progress in a negative direction is monitored by high-tech scans, photos and self-administered tests of the eyes by having patients look at an Amsler grid daily.

The "Hope" chapter of this book (page 10) told of changes in Mark's vision that occurred during his first two months of weekly IVs. Because only negative change is expected, he had difficulty believing that his vision could possibly be improving. "I must be looking at things more closely and that's why I'm seeing this color," he often said. However, when some yellow objects such as traffic lights and lines down the center of roads began to appear orangish-yellow to him, he suddenly remembered a school-bus yellow truck he had spent hours playing with before he was kindergarten age. The realization that orangish-yellow *was* and still *is* the color that warns us "Be careful!" made the improvement in his vision seem more real.

Five months after he received the diagnosis of macular degeneration, we attended a concert. He decided to wear his old glasses because he was wearing dress clothes. His new glasses, made using the prescription he got at the eye exam where he first heard "You have macular degeneration," have safety side shields which he did not want to remove to make them suitable to wear to the concert.

Wearing the old glasses was "a shock," he said. He had been attributing improved ease of seeing his computer monitor at work to having a new prescription in his glasses. The evening of the concert, he was amazed that he could see very well with his old glasses. The change in his vision was not due to the new prescription: it was his eyes that had improved.

At that time, he had taken twelve IVs. He also had just had an appointment with a holistic eye doctor recommended by our dentist. The new eye doctor had a special camera to photograph the retina. She took photographs of his retinas and compared them to photographs taken by the retina specialist over four months previously.

"No change," she said, which is usually the best news possible for a macular degeneration patient. She also told us that she had been looking at retina photos for over thirty years.

The allergy doctor who many years previously told my mother about the macular degeneration IV protocol "stopping it in its tracks" was very pleased to hear about the improvements in vision Mark had noticed and that the retina photos showed no change. He recommended that Mark take the IVs every other week until his six-month appointment with the retina specialist. After that, if the report was also positive, he said Mark could take the IVs at monthly or longer intervals, but he should be taking zinc, selenium and taurine supplements daily. (See page 27 for more about these supplements including dosages). He also recommended that Mark have a self-test[1] that would detect the beginning of possible deterioration in his vision. He said a self-test would reveal any deterioration in vision before the scans, photographs and examinations by eye doctors could detect the change.

As the six-month appointment with the retina specialist grew near, we both began to dread it because the first appointment had been so depressing. Mark's goal was to remain on good terms with the doctor so if he ever experienced a crisis and needed an anti-VEGF injection, she would take care of him quickly. Since she seemed very conventional, he decided not to tell her about taking IVs, etc.

On the day of the appointment, he first saw the same young assistant as on the previous visit. As part of taking Mark's recent history, he asked, "Have you noticed any changes in your vision?" Mark replied that he was seeing blue and yellow better and that he could see his computer monitor at work more easily. "But I did have a refraction before I first came here in December and got new glasses in January," he explained. However, Mark had disproved the "new glasses made the difference" theory to himself when he wore his old glasses to drive to the concert a month previously.

Our next stop was the mid-office waiting area. I whispered to Mark that I was surprised that he had mentioned seeing blue and yellow better and asked, "What are you going to tell the doctor if she asks you about why the improvement happened?" Before he could reply, a technician summoned him for eye scans and photographs.

When the tests were finished, we were ushered into the room where the doctor would examine his eyes and discuss the test results and her findings with us. After the examination, she looked at his test results on a computer monitor. Her back was to me and I could see the monitor. She studied the OCT (Optical Coherence Tomography) scan quietly. She asked Mark if he had ever smoked. He replied, "No, but my parents did and I got plenty of second-hand smoke growing up."

Then she examined a set of four color photos of Mark's retinas, each of which was a composite of photos taken from several different angles. She began making "Hmmm... hmm... hmmm... hmm... hmm" sounds as she clicked with her computer mouse. What was she seeing? The "hmm" sounds drove me to the verge of panic. I looked closely at the photos to try to determine what she saw. They all looked very similar aside from slight variations in color. The two upper images on the monitor were bright red with a small yellow spot in the middle of the cloud of orbs. The two lower images also contained a yellow spot (the macula) in the center and but were duller red and even grayish-red in some places.

She turned her chair to face him squarely and said, "Your macular degeneration is *definitely* stable. Whatever you're doing, *keep doing it!*" Then she replied to Mark's comment about his parents smoking and she told us about her childhood

---

1 Mark's self-test is paying attention to how easily he can see and work at his computer monitor at work. Before he began taking the IVs, he had to take his glasses off and put his nose close to the monitor, put them on, take them off, put them on, etc. all the time. If he ever cannot keep his eyes at a set distance from his monitor and work comfortably with his glasses on, he will know it is time to return to weekly IV treatments until his vision again improves. The doctor thought this sounded like a good self-test.

experiences with smoking. She described growing up on a farm in Iowa, her father smoking, giving her a puff of his cigarette when she was about ten years old, and how that experience "cured" her of smoking for life. However, her three siblings, who did not take a puff at an early age, all smoke now. She told us she'd recently found her bronzed baby shoes in her basement. They were part of an ashtray set and she had dropped the ashtray. Mark later said she was like a totally different person than she had been on our first visit. She seemed human instead of an aloof authority figure.

The next day, as I was mentally processing what had transpired at the appointment, I remembered having read that the color of a normal, healthy retina is bright red. However, with macular degeneration the color dulls and can become mottled.[2] Perhaps a change to a more normal retina color was what she had seen! I also realized that it must be very difficult for a doctor to have to tell patients terribly depressing news for most of the day every weekday. When she saw better-than-usual retina photos, she might have been so surprised that she lost her inhibitions and opened up to us. As we shared baby shoe and other stories, she became a real person under her white coat. By the end of our visit, we felt that we had begun to know her personally and she was learning about us. Without knowing why she saw what she saw, she gave us comfort.

Although she did not know what we had been doing, much of the "doing" happened in the kitchen where the next several chapters will take us. These chapters will make cooking easier than you may expect. Time in the kitchen will improve your nutritional status which is the most effective factor for improving outcomes with macular degeneration.

---

2 This information is on page 22.

Putting Superior Eye Nutrition Into Practice

Hippocrates said, "Let your food be your medicine and your medicine be your food." Although the naturopath Mark consulted for macular degeneration gave him IVs and supplements, she said "food is best" repeatedly. The purpose of the next several chapters is to help macular degeneration patients and their families improve eye health where superior eye nutrition begins – in the kitchen.

The stakes are high! By shopping carefully and preparing food at home, you gain control over your nutritional status which is essential for preserving and improving vision. If you consider yourself a non-cook, now is the time to exercise skills you already have (using a knife, plugging in an appliance, pushing buttons) and prepare the most important recipe in this book, the "Eye Smoothie" on page 87. *Now* you will begin doing what most needs to be done to preserve your sight.

At every appointment, the retina specialist has stressed the importance of consuming sufficient quantities of foods from the list she gave him. Topping the list is kale, with 21,900 micrograms of lutein per 100 gram (3.5 ounce) serving of raw kale. Like many people, Mark doesn't like kale. Yet he was told to eat one cup cooked or four cups raw of dark green leafy vegetables including kale every day.

Mark ate raw spinach willingly at first but then became so tired of it that he'd pick out the spinach from his salad as soon as he arrived from work to eat it and "get it over with." An old friend who eats nutritiously sympathized, "I don't blame Mark for getting tired of spinach. A little is OK, but I prefer a mixture of greens containing some spinach. Spinach in pizza, lasagna, pasta, omelets is OK, but cooked, alone... 'No, thanks.' I don't like kale either, too bitter."

Thankfully, the naturopath told me how to get the prescribed 100 grams of kale into him every day in a smoothie. (I purchase about one and a half pounds of kale per week, so I know he's getting his quota). The smoothie made using her suggestions gives him kale and spinach for lutein and zeaxanthin, omega-3 fatty acids, blueberries for anthocyanins, and coconut milk with fat that improves the absorption of the fat-soluble nutrients. If you follow only one piece of advice from this book, let it be to consume an eye smoothie every day. (The recipe is on page 87).

Shopping and Stocking the Pantry and Refrigerator

The best way to begin cooking is with all the needed foods and ingredients on hand. The foods listed below should always be in your home or, as in the case of perishable foods such as fish and vegetables, on your grocery list every week. This list of foods to keep on hand is based, first and foremost, on what is most likely to preserve and improve your vision. As you begin stocking your pantry, discard or donate everything your find there that is harmful to eyes.

Although no one I know has a traditional pantry (room off the kitchen staffed by a butler), many are blessed with a narrow closet near the kitchen, storage shelves in the basement, or a special kitchen cabinet where favorite and/or essential room-temperature stable foods are stored. With a well-stocked food storage area, when we can't get out to shop, we can still eat well. A pantry gives us comfort and security; food in the pantry is like money in the bank.

Cold Water Fish

Salmon, mackerel, sardines or tuna should be eaten at least weekly, says Mark's retina specialist but, in Save *Your Sight!* ophthalmologists Marc and Michael Rose, MD, recommend having cold-water fish two or three times a week. Wild caught fish provide omega-3 fatty acids and vitamin A in a form that is more easily absorbed and utilized than beta-carotene from plant sources. The vitamin A is used to make eye-protecting pigments and any excess is stored in the liver for future use.[1] Salmon also contains astaxanthin. (See pages 28 to 29 for more about this super-antioxidant). Wild-caught salmon provides greater anti-oxidant activity than farmed salmon because wild salmon consumes a natural source of astaxanthin, the tiny marine crustaceans krill. **Because farmed salmon eats an unnatural diet, it is high in pro-inflammatory omega-6 fats.** Read more about this on pages 38 to 39. Although farmed salmon is more affordable, either eat a smaller portion of fresh wild-caught salmon or eat canned wild-caught salmon for best eye health.

Although fish is tastiest when fresh, you may want to keep canned fish on hand in case time constraints or circumstances prevent grocery shopping. A delicious way to prepare canned wild-caught salmon is "Salmon Burgers" on page 100. Here in Colorado, far from the ocean, fish frozen on the boat immediately after being caught tastes best at times, so I also keep some frozen fish in our freezer.

Vegetables High in Lutein and Zeaxanthin

Stock your refrigerator at least weekly with dark green leafy vegetables containing high levels of lutein and zeaxanthin. The retina specialist's list of prescribed foods listed recommended vegetables in order of the amount of these pigments they contain. She wrote "daily" by those containing greater than 10,000 micrograms per 100 gram serving, which are kale, collards, spinach and fresh parsley. For a similar printable list containing an assortment of foods, visit this webpage: https://www.macular.org/wp-content/uploads/2016/05/lutein.pdf . For a list of vegetables high in lutein and zeaxanthin that is much more complete than the retina specialist's, go to this webpage: http://foodinfo.us/SourcesUnabridged.aspx?Nutr_No=338 .

Make salads using raw spinach for most or at least a large part of the vegetables. A recipe for cooking any of the leafy greens is on page 113. Try cooking kale with

---

1 Rose, Marc R., MD and Michael R. Rose, MD. Save *Your Sight! Natural Ways to Prevent and Reverse Macular Degeneration.* (New York, NY: Warner Books, 1998), 73.

bacon for improved flavor using the recipe on page 114. Kale chips are tasty and come in a wide variety of flavors. See "Sources," page 218 for where to purchase the brand of kale chips Mark ate. If you wish to make kale chips, a recipe is on page 162. Parsley or spinach pesto is a wonderful condiment spread on bread or crackers and delicious on pasta or spaghetti squash, especially if topped with grated Romano cheese. The recipe for both types of pesto is on page 165.

Colorful Vegetables

Colorful red, orange, yellow and purple vegetables are high in carotenoids and should be added to every vegetable repertoire and weekly grocery list. Use tomatoes and red, orange or yellow bell peppers in salads. Purple cabbage is delicious in coleslaw (page 120) or roasted in the oven (recipe on page 117). Yellow or orange winter squash and sweet potatoes can be baked or, using the recipes on pages 118 to 119, made into special crisp vegetable treats. Eat cooked tomatoes as part of a fat-containing meal to get the carotenoid lycopene which combats macular degeneration. For more about lycopene, see page 38.

Fruit

Fresh fruit is loaded with antioxidants and other nutrients and, especially in season, is a real treat, a good low glycemic index snack and a replacement for common dessert foods. Purchase dark red and purple fruits and berries for their anthocyanins fresh when possible and affordable. Use frozen berries when fresh are out of season. For the "Eye Smoothie," page 87, solidly frozen blueberries are required, so keep your freezer well stocked with frozen blueberries.

Eggs

Egg yolks are high in lutein in a form that is easily absorbed and utilized, so if you are not allergic to eggs, put them on your grocery list. Because macular degeneration patients eat eggs for the lutein, consider purchasing eggs from naturally raised chickens. They have deep yellow yolks (a color which says "more lutein" to me) and are often featured in health food store sales. Mark's retina specialist recommends eating four eggs or egg yolks per week.

However, eggs are quite allergenic and so are not for everyone. Allergic reactions should be avoided to prevent fueling eye-damaging inflammation. Mark tolerated eggs well until he'd eaten them once a week for about five months. Then he stopped enjoying eating them in "Eggs Benedict" (recipe on page 96) or a frittata (page 95) and wanted them hard boiled only. Then, they gave him hives. He waited six months before trying them again and now tolerates them in the small amounts used in baking. For more about eggs, see page 34.

Pantry Staples

A good way to begin eating for eye health is by ridding your home of all foods that can damage the eyes. Remove and donate or discard all cookies, candy, crackers (except for whole grain varieties such as Wasa™ or Ak-Mak™ crackers), chips, white flour, white rice, cold cereals (except for 100% wheat bran), quick or instant oatmeal, baking mixes, sugar, corn syrup including pancake syrup made from it, artificial sweeteners, jelly and jam unless sweetened with only fruit sweeteners, canned goods containing sugar or corn syrup, canned goods containing MSG or similar taste "enhancing" chemicals including canned soup and broth, and all unhealthy fats such as traditional shortening, corn, soy, and safflower oil. Then re-stock your panty with the wholesome eye-health-promoting foods below.

Grains and Foods Made From Grains

Stock your cupboards with whole grains such as brown rice and other less common whole grains to cook as side dishes. Quinoa is higher in protein and contains the most complete amino acid profile of any grain. To introduce your family to alternative whole grains, try "Quinoa Stuffed Peppers" (recipe on page 102) or "Quinoa Pilaf" (page 125). Then expand your whole grain side dish choices to include wild rice, buckwheat, teff, amaranth and more. Recipes for cooking these whole grains are on pages 122 to 125.

Purchase whole grain flours, preferably stone-ground, such as Bob's Red Mill™ stone ground whole wheat and quinoa flours. Rice flour, even brown rice flour, is high glycemic and should be avoided if you have any blood sugar issues. All baking mixes should be avoided. For breakfast, try thick-rolled oatmeal (see the easy overnight recipe on page 93), or better yet, have a high-protein breakfast.

Healthy Fats and Other Cooking and Baking Ingredients

Purchase monounsaturated oils such as olive, avocado and organic non-GMO canola oil. (Read more about oils and the canola controversy on pages 37 to 38). Spectrum Naturals™ Shortening, which is non-hydrogenated organic palm oil, will replace traditional shortening in any recipe and makes very flakey pie crusts. If you plan to bake, you may want to stock leavening ingredients such as corn and aluminum-free baking powder, baking soda, lemon juice and, if you are allergic, unbuffered vitamin C powder or crystals. See "Sources," pages 212 and 217 for where to get these leavening ingredients.

Yeast is a leavening ingredient that is stored in the freezer and/or refrigerator rather than the pantry. For where to get yeast in economical one pound bags, see "Sources," page 217. Store the bags, unopened or opened and taped closed, in the freezer. Occasionally remove the bag and transfer enough yeast to last a month or so into a glass jar. Store this jar in the refrigerator, ready to use whenever needed.

Sweeteners

For the sake of your eyes, make plant-derived non-nutritive sweeteners such as enzyme-treated stevia extract and monk fruit extract your sweeteners of choice and stock them in your pantry. They are used in every recipe that requires sweetening in this book. Use them in beverages and substitute them for other sweeteners in recipes not from this book as well. To substitute them for sugar, a very approximate rule of thumb is to use one teaspoon of stevia or monk fruit extract to replace a cup of sugar.

For more help with substitution, visit this webpage: https://www.wholesomeyum.com/natural-low-carb-sweeteners-guide-conversion-chart/ and scroll down to the last fifth of the page to see a chart of conversions between sugar and stevia or monk fruit. If you sign up for the Wholesome Yum emails, you will be able to download a printable version of this chart and will get a code for 20% off on Lakanto™ pure monk fruit extract. This code may be used repeatedly and does not expire. However, don't let the Wholesome Yum emails convince you to bake with erythritol. It is made from corn, which I assume is GMO corn because I have not seen organic erythritol or any that bears the non-GMO logo. Additionally, it may produce side effects which include diarrhea, headache, and stomachache in some people. Because pure monk fruit extract is pricey, it is often cut with erythritol or other sweeteners.

You may also want to purchase some agave syrup to use when others beside your family will be eating what you make. Agave is a very low glycemic index sweetener which does not have any unusual flavor. Because of its low glycemic index, it may be consumed in small amounts on special occasions such as when you have guests.

For more about all of these sweeteners, see pages 82 to 84. See "Sources," pages 216 to 217 for where to order them.

Snacks

Raw nuts and seeds are nutrient-rich blood sugar stabilizing snacks. Purchase several varieties and keep them on hand at all times for good protein snacks. Brazil nuts are one of the best food sources of selenium, a trace mineral which is important for eye health. If you have trouble digesting nuts and seeds, see the recipes on pages 162 and 163 for how to make them more digestible. See "Sources," pages 218 to 219, for where to purchase easily digested nuts and seeds and nut or seed butter made from sprouted nuts or seeds.

Whole grain crackers such as Ak-Mak™ (100% whole wheat) or Wasa™ (100% whole rye) crackers make good snacks with natural nut butter. Be sure to purchase nut butter without added sweeteners or fats, especially hydrogenated fats.

I enjoy snacking on the sea salt flavor "Crunchy Chick Peas" made by Saffron Road™. Your health food store may carry them in a variety of flavors as well as selling kale chips made in many flavors. See "Sources," pages 218 to 219, for where to order any of these snacks if you cannot find them locally.

Canned Goods

Although fresh fruit contains more nutrients, you may wish to keep some canned fruit such as pineapple in its own juice on hand. All fruit should be water or juice packed, not sweetened with sugar or corn sweeteners. As mentioned on page 74, canned fish, especially eye-health-promoting salmon, tuna, mackerel or sardines, is good to have on hand for times when you can't get out. The only canned goods I use routinely are tomato products. Make sure they do not contain sugar, corn syrup, or other ingredients. Avoid all canned goods that contain MSG in any of its forms (see page 201 for names it is disguised by) or other flavor "enhancing" ingredients.

Meal Makers

Stock your pantry with a few foods that can make a meal in a pinch. Purchase frozen bone broth[2] (See "Sources," pages 212 and 213 for where to get it) and keep it in your freezer to replace the artificial ingredient-laden canned broth or soup formerly in the pantry. Leftovers and bone broth can be used to produce a quick soup. For example, combine chicken broth with leftover cubed chicken, leftover peas, beans and/or other vegetables and cooked brown rice or quinoa from last night's dinner for a light meal. Dry legumes are the beginning of delicious soups (pages 103 to 107), crockpot baked beans (page 116), chili (page 108) and other meals.

Ingredients for High-Nutrient Sweet Treats

Everyone needs a treat occasionally, so plan ahead. Although cacao candy should be refrigerated, the ingredients for making it can be kept on hand at room temperature. For the maximum nutrient and mood boost from cacao candy, purchase organic cacao powder and cocoa butter. Navitas™ cocoa butter comes in bags of chunks which are very convenient for measuring.

For a sweet ready-to-eat snack, stock goji berries which contain zeaxanthin, beta-carotene, lycopene, and many vitamins and minerals, including trace minerals, that support eye health.[3] See the recipe for "Goji GORP," page 161, which contains protein-rich nuts to balance the carbohydrate in the berries. Store the GORP in the pantry for a quick-to-grab, protein-balanced sweet treat.

Now that you are well stocked, turn to the next chapter for more about how to make these foods and ingredients into delicious meals that will promote eye health.

---

2 For bone broth recipes, refer to *Healing Basics*, described on page 327.

3 Degner, Leslie, RN, BSN. "Goji Berries Benefit to the Macula." *Macular Degeneration News.* http://www.webrn-maculardegeneration.com/goji-berries-benefits.html

About the Recipes and Ingredients

The recipes in this book are unique. They have been selected and developed to contain the foods and nutrients that are most beneficial for eye health. I also thought about the cook when developing these recipes. Some cooks might be macular degeneration patients themselves with visual challenges. I realized that I could not begin to comprehend how difficult tasks such as cooking may be for some macular degeneration patients. For lack of other resources, I Googled "cooking with low vision" and discovered that there are special cooking devices (even talking thermometers to tell when the meat is done[1]), eccentric viewing, and low vision rehabilitation specialists who may be able to help. On the VisionAware website[2] I found comments from low vision cooks who had derived great satisfaction from cooking for themselves and living independently. Therefore, this book contains some recipes that have been adapted for use by patients themselves. These and other recipes will be in the planned large print edition of this book; most complex recipes will be omitted.

Other cooks may be family members and friends of macular degeneration patients. Our loved ones experienced a devastating blow when they were diagnosed with a condition that threatens to destroy their sight, goals, dreams and almost everything that they currently do or hoped to do in the future. They are experiencing grief and loss, and some of the loss they feel may occur when they want to eat and find their choices dictated by orders from eye doctors.

Family members and friends have more power to help these loved ones than you might imagine. With superior eye nutrition, their sight and dreams may be preserved. However, the stabilization and improvement will take time. For immediate, tangible comfort today, this book includes recipes for foods that say, "Pleasure in eating is not a thing of the past," and, "I care."

The recipes in this book are also unique because they can be used by those on diets restricted by gluten intolerance, food allergies or diabetes. Except for a few wheat-containing breads, the recipes are allergy-friendly and gluten-free. For diabetics, the recipes list the number of carbohydrate units per serving at the end of the recipe.[3] Some recipes also list the number of protein units per serving[3] so those on the insulin resistance diet can "link and balance" carbohydrates and proteins in their meals and snacks. See pages 43 to 46 and 177 to 182 for more about how to keep

---

1 Maxi-Aids is a source for independence devices that my aunt used when she was at home. See this page for talking cooking thermometers: https://www.maxiaids.com/talking-thermo meters?gclid=CjwKCAjw2rjcBRBuEiwAheKeL1toI0AKaPNHqVYiTzqi6kFOVO2nOg_ wZqjuUwlfv2iBiBZhU0g13RoCnmIQAvD_BwE . Better lighting may help with many types of tasks: https://www.webrn-maculardegeneration.com/low-vision-lighting.html .

2 " Ms. Dorothy Keeps on Cooking" http://www.visionaware.org/info/emotional-support/personal-stories/living-and-coping-with-vision-loss/ms-dorothy-keeps-on-cooking/1235 ; "Two Blind Cooks in my Kitchen," https://www.visionaware.org/blog/visually-impaired-now-what/two-blind-cooks-in-my-kitchen/12

3 Each carbohydrate unit is 15 grams of carbohydrate; each protein unit is 7 grams of protein.

blood sugar stable and about using the insulin resistance diet for weight loss without experiencing the hunger that is so common on low-fat and low-calorie diets.

Although it has been a goal to *not* over-complicate a recipe or the cook's life, some more challenging recipes that were helpful for Mark are included here such as sourdough bread.[4] Your loved one may not really need these recipes.

Realism must be a part of the recipe sections of this book. Cooking is time-consuming work, but it does not have to be as much work as it may seem the first time you read the second half of this book. I suspect that most macular degeneration patients are not semi-gourmets like Mark and will not need a wide variety of menu choices to be happy on an eye-healthy diet. **You don't have to do it all.**

The basics for saving or improving vision include:

(1) Every macular degeneration patient should **have an "Eye Smoothie,"** page 87, **every day** and

(2) Should consistently follow the advice of ophthalmologists Marc and Michael Rose, MD, to **eliminate all sugar and processed foods.** These changes may be enough to stabilize blood sugar levels for many patients.

(3) In addition to good diet, macular degeneration patients should **take supplements that supply the nutrients listed on page 27 to 28** and

(4) Avoid prescription and over-the-counter medications that can cause macular degeneration to progress quickly. Read more about these drugs on pages 53 to 58.

(5) Finally, macular degeneration patients should **eat salmon or other cold water fish weekly** or more often. An easy recipe for baked salmon is on page 99.

There are many recipes here if you want to use a few of them often and most occasionally, but you do not need to become compulsive about cooking. Following the advice in bold above should make a difference hopefully without being an excessive burden. The Drs. Rose encourage moderation in all things including what is eaten.

To stick with any type of diet for all of your remaining years, it must be livable and satisfy hunger and appetite. It must also provide some pleasure as well as plentifully supplying the nutrients required to stop the progression of or reverse macular degeneration. I hope the recipes in this book make following an eye-healthy diet possible long-term and enjoyable and that this book will be comforting to both the patient and the cook.

Ingredients

Some of the recipes contain ingredients which may be unfamiliar. Therefore, the less commonly used ingredients are discussed on the following pages.

---

4 Although the sourdough bread recipes take more time than other recipes, they can be made more easily than you might expect with the right bread machine and ingredients such as a freeze-dried *Lactobacillus* culture instead of maintaining a traditional sourdough starter. See pages 137 to 138 for more about this less laborious way of making sourdough bread.

Grains and Flours

When Mark's naturopath said, "No white rice," she mentioned quinoa as a possible alternative. Quinoa is not a true grain. It is the seed of a plant in the same family as spinach, beets, and Swiss chard. Nutritionally superior, it is a very versatile gluten-free replacement for other grains and flours.

Most of the recipes in this book are food-allergy friendly and gluten- free. This allows macular degeneration patients with special dietary needs to avoid eating problem foods which can stimulate eye-damaging inflammation. Therefore, you will see recipes using unfamiliar grains such as gluten-containing spelt and gluten-free amaranth, quinoa, buckwheat, and teff. Starches such as arrowroot and tapioca flour are used to make some of these flours more cohesive. Nut flours can produce baked goods which contain no carbohydrates, a great help to diabetics and those on glycemic control diets. An introduction to less common grains and flours is below

QUINOA boasts a good content of high-quality protein. It is one of the best protein sources, due to its amino acid balance, among plant foods. For this reason, it is very satisfying to eat; quinoa baked goods really stick with you. Those allergic to all true grains will find quinoa a welcome dietary staple. It has a distinctive taste so is best used with other strongly flavored ingredients such as cinnamon, sesame seeds, cacao and carob. Quinoa flour is excellent in baked goods of all kinds and makes good yeast bread. Whole grain quinoa has a natural soapy coating on it, so before you cook it, put it in a strainer and rinse it under running water until the water is no longer sudsy. This coating protects the plant from insects.

AMARANTH is in the same botanical order as quinoa, although it is not in the same food family. Because it is not in the grain family, it is a welcome dietary addition for those allergic to all true grains. It makes very tasty baked goods. Purchase it at a store that refrigerates its flour and refrigerate or freeze it at home since it may develop a strong flavor if stored too long at room temperature.

BUCKWHEAT is a tasty whole grain and makes versatile non-grain flour. It is excellent in waffles and pancakes. In a few recipes (such crackers and muffins, which are not included in this book) it can have a strong flavor. It is not a true grain but is in the same food family as rhubarb.

TEFF is a gluten-free true grain. It has been difficult to find in the past, but now Bob's Red Mill™ has made it easily available. It is less bland than the other gluten-free grains but is still delicious. Teff flour tends to be a little gritty but makes great baked goods.

SPELT is closely related to wheat with its biological classification being *Triticum spelta* while wheat is *Triticum aestivum*. In spite of the close relationship, spelt is tolerated by many wheat-allergic individuals. Spelt makes excellent yeast bread because it is as high in gluten as wheat although its gluten is more soluble and may be more easily digested for this reason.

The spelt recipes in this book were developed using Purity Foods™ flour. Purity Foods mills its flour from a European strain of spelt that is higher in protein than most spelt flours. so it is excellent for yeast bread. Shortly after I first began baking with spelt, other companies began producing spelt flour. These other flours did not work consistently for non-yeast baking and were unacceptable for yeast breads, so I kept using only Purity Foods™ flour for over two decades.

Recently I visited Purity Foods' "Nature's Legacy" website to purchase spelt pasta and discovered that many products were out of stock. I hope this is a temporary shortage, but it might be the result of a FDA policy that could cause the demise of the company.[5] Therefore, I purchased Arrowhead Mills™ and Bob's Red Mill™ whole spelt flour and tested them in the spelt recipes in this book. Both worked well for non-yeast baking. The Arrowhead Mills™ was also excellent in yeast bread (page 133). The See "Sources of Special Foods," page 215, for where to get spelt flour.

ARROWROOT and **TAPIOCA STARCH** serve as binders in gluten-free baking and can also be used as thickeners for sauces and gravies. Arrowroot and tapioca starch can be substituted for each other in baking in equal quantities.

NUT FLOURS contain no carbohydrates but are a good source of protein, healthy fat, vitamins and minerals. **Almond flour** has the added advantage of being available in a very finely ground form that is good for baking. Not all almond flour is suitable for baking however. It must be made from blanched (skinless) almonds and be very finely ground to produce consistently acceptable baking results. In addition, nut flours should be stored in the refrigerator or freezer to keep the fat from becoming rancid. Of the types of finely-ground blanched almond flour available, I prefer Honeyville™ almond flour because it is economical and excellent for baking. See "Sources," page 214, to purchase this flour.

Sweeteners

This book uses stevia and monk fruit extracts in most of the recipes which require a sweetener. Both are natural, plant-derived, non-nutritive sweeteners and

---

5 For the last several years, the US government has required foods to be labeled to indicate whether they contain any of eight food allergens. As part of this policy, the FDA declared that spelt is wheat and spelt products must be labeled "Contains wheat." Although spelt and wheat are indeed closely related, they are two different species in the same genus. When asked why they ruled that spelt is wheat, a FDA official said that it was because spelt contains gluten. (They had no answer to the question of whether rye should also be considered wheat because it contains gluten).

Spelt does contain gluten and should not be eaten by anyone who is gluten-sensitive or has celiac disease, but the presence of gluten does not make spelt wheat. The gluten in spelt behaves differently than the gluten in wheat in cooking. It is difficult to make seitan from spelt because its gluten dissolves in hot water. Because the gluten in spelt is more soluble than wheat gluten, making yeast bread with spelt is also different from making it with wheat. There are some bread machines that work quite well for wheat and even other allergy breads but are unacceptable for spelt bread because they knead so vigorously that they over-develop the gluten.

It is possible that the greater solubility of spelt protein makes it easier to digest than wheat, or more likely that spelt is better tolerated than wheat because it has not been geneticaly manipulated. Undoubtedly, most people have had much less prior exposure to spelt than to wheat, resulting in less opportunity to become allergic to spelt. Whatever the reason, there are people who suffer allergic reactions after eating wheat but do not react to spelt. I have talked to many of them. Review **your** food allergy test results and follow the diet recommended for you by your doctor, but there is no need to restrict spelt consumption based on error-ridden government labeling requirements.

have no effect on blood sugar. Agave is a natural nutritive sweetener with a very low glycemic index score and is given as an option in some recipes.

STEVIA comes from an herb which is promoted as a supplement helpful for diabetics. This herb, *Stevia rebaudiana,* is a member of the composite (lettuce) family. Stevia has been used as a sweetener in Japan, Paraguay, Brazil, and other parts of the world for hundreds of years. It is available in several forms such as finely crushed leaves, liquid, and dry extract powder.

Most stevia sold in grocery or health food stores has a licorice-like taste which may be noticeable in bland recipes but is almost undetectable in recipes containing strongly flavored ingredients. However, several years ago a new more neutral tasting white stevia powder was developed which is treated during production with an enzyme that reduces the licorice taste. Some brands of this "next generation" stevia powder are cut with allergenic ingredients such as dextrose or maltodextrin, which usually comes from corn, or other fillers. The stevia recipes in this book were developed using the most neutral, purely-sweet tasting of the new stevia powders which contains no fillers and also is the most economical. It may be ordered from Berlin Seeds.™ See "Sources, page 216, for ordering information.

MONK FRUIT comes from a melon tree native to Asia, *Siraitia grosevenorii.* Because the sweet molecules in the melon (mogrisides) are present in only 1% concentration, considerable refining is needed to produce monk fruit extract. For this reason, it is more expensive than stevia extract.

The taste of monk fruit in low concentrations is pure sweetness. In higher amounts, it can have an off taste described as "old squash" on the internet and "weird" by my son Joel. Because of the higher price of monk fruit extract, it often is cut with other sweeteners such as dextrose, erythritol or other sugar alcohols. The monk fruit used in the recipes in this book is pure monk fruit extract with no other ingredients added. See "Sources, page 217, for where to purchase pure monk fruit extract. The brands that I have used are Pure Monk™, Lakanto™ (30% mogrisides), and NuNaturals™. They seem interchangeable; I've used all three brands in the same amounts in the recipes here with good results.

My tasters have noticed a difference between stevia and monk fruit that affects how much they like a sweetener in some recipes. The two sweeteners have different time patterns of how quickly the tongue perceives "sweet" and how long the sweet taste lasts. Stevia tastes sweet slowly, but the sweetness lasts. Monk fruit tastes sweet immediately but the sweetness does not last long.[6]

Therefore, they prefer stevia in "Sugar-free Cola," page 91, because the beverage passes through the mouth quickly and is most enjoyable if the sweet flavor lasts as long as the flavors of cola, vanilla, etc.

---

6 For the best of both worlds with these sweeteners, a taster suggested using both monk fruit and stevia together in a recipe, with monk fruit as about ⅔ of the sweetener and stevia as the remaining ⅓. This change improved the perception of "sweet." However, I recommend that only those who are totally non-allergic and would like more persistent sweetness try this. If you have allergic tendencies, I'd advise not eating either sweetener every day; rotate them to maintain tolerance.

For "Cacao Candy," page 172, they prefer monk fruit because cacao is quite bitter. (The bitter taste comes from polyphenols which produce most of the health benefits of cacao).[7] If the bitterness of cacao is tasted immediately with no sweetness to balance it, it's not pleasant, so monk fruit is the preferred sweetener because its sweetness impacts immediately.

Stevia and monk fruit are good sweeteners for cooking as well as for health. Unlike some man-made non-nutritive sweeteners, stevia and monk fruit can be heated to any cooking temperature and for any length of time without breaking down. Expect cookies to be light colored; baked goods made with non-nutritive sweeteners do not brown much. The amount of stevia or monk fruit in the recipes is usually given as a range because the perception of how sweet they taste varies from person to person. In addition, for foods that contain a large amount of either of these sweeteners, how much a person detects the licorice or squash flavor varies from person to person. The longer you use stevia, the less you detect the licorice-like flavor and the more you are likely to want in a recipe. Until you know how much stevia you prefer, use the smaller amount in the recipe the first time you make it. Then increase the amount gradually with successive batches of the recipe until it tastes best to you. If you prefer more than the largest amount given in the recipe, that is fine. It will not affect how the recipe rises, etc.

Both stevia and monk fruit extract powders are used in very small amounts in some recipes. To measure them easily, see "Sources," page 211 for measuring spoons that can measure amounts such as one-third and one-sixteenth of a teaspoon.

The bottom line on stevia and monk fruit is – they are both great! They send cravings for sweet foods running and satisfy the desire for something sweet without affecting blood sugar or leading to weight gain. Having treats sweetened with stevia or monk fruit can make a difference in how contented one is with life and compensate for some of the foods given up and other challenges faced as you make changes for the sake of your eyes.

The final sweetener that you may want to use when you have guests is **AGAVE**. It is natural and plant-derived but nutritive, although it has a very low GI score. It is made from a succulent native to Mexico, *Agave americana,* or other species of *Agave*. Its taste is purely sweet with no aftertaste or unusual flavor. If used in place of sugar, use about two-thirds as much agave as sugar. Also, decrease the liquid in the recipe to compensate for the liquid contributed by the agave.

Cacao and Carob

When I first saw the word "cacao" in a magazine from the health food store, it was used in a recipe as an ingredient that seemed to be interchangeable with cocoa, so I didn't pay much attention. Then I read about it in *What You Must Know About Age-*

7 Mercola, Dr. Joseph, DO. "Dark Chocolate Reduces Stress and Inflammation, Boosts Memory and Mood." May 10, 2018. https://articles.mercola.com/sites/articles/archive/2018/05/10/dark-chocolate-benefits.aspx

Related Macular Degeneration by Dr. Jeffery Anshel, OD, and Laura Stevens. Ms. Stevens has macular degeneration herself and included a chapter in the book about foods and supplements that have improved her vision. She wrote that while studies have not been done specifically on the effect of chocolate on macular degeneration, there are studies that show that consumption of dark chocolate helps risk factors for macular degeneration such as insulin resistance and high blood pressure.[8]

CACAO is very different nutritionally from cocoa and from most commercially made chocolate. The word "cacao" refers to the plant and its raw unroasted beans. Raw cacao beans are extremely high in antioxidants, phenolic phytochemicals, flavonoids and many minerals, especially magnesium.[9] The high-heat roasting and heavy processing that turns cacao beans into commercially made chocolate destroys much of their nutritional value. The addition of milk further reduces the nutritional value because dairy products interfere with our ability to absorb phytonutrients. Therefore, the milk chocolate most commonly eaten provides few nutrients and the sugar it contains promotes blood sugar and weight problems.[10]

Cocoa is the ground solids from cacao beans that have been roasted. Cocoa butter is the fat pressed, before roasting, from the beans. Cacao powder is the ground solids from raw, unroasted beans and retains the full amount of original bean nutrients. Cacao powder can combined with cocoa butter and stevia or monk fruit extract to make an extremely nutritious sugar-free treat. In addition to enjoying the flavor of cacao candy, Mark definitely experiences a mood-elevating benefit when he eats it. See the recipe on page 172.

CAROB powder is ground from the seeds of the locust bean and is in the legume family. It is usually used as a chocolate substitute. Carob powder is naturally sweet, so may require less sweetener in recipes than cacao does or even no sweetener. Carob chips are a welcome addition to cookies for those who are allergic to chocolate. Although carob does not offer the nutritional benefits of cacao, it can keep your allergic loved one from being deprived of treats and thus help them consistently follow an eye health diet. A recipe for carob chips and candy is found on page 170.

Both cacao and carob powder tend to form lumps upon standing, so you may need to press them through a wire mesh strainer with the back of a spoon before using them in a recipe. I discovered an old flour sifter in my basement and found that using it to remove the lumps makes this task go much more quickly and easily.

Now that you've become acquainted with the less common ingredients in this book, you are ready to begin putting superior nutrition into practice in the kitchen. You can do it! It is THE BEST way to preserve and improve eyesight. Continue reading for recipes.

---

8 Anshel, Jeffrey, OD and Laura Stevens, M. Sci. *What You Must Know About Age-Related Macular Degeneration.* (Garden City Park, NY: Square One Publishers, 2018), 73-76.

9 Mercola, Dr. Joseph, DO. "What Is Cacao Good For?" https://foodfacts.mercola.com/cacao.html and Mercola, Dr. Joseph, DO. "Dark Chocolate Reduces Stress and Inflammation, Boosts Memory and Mood." May 10, 2018. https://articles.mercola.com/sites/articles/archive/2018/05/10/dark-chocolate-benefits.aspx

10 Pemberton, Corey. "Cacao vs. Cocoa: The Difference and Why It Matters." August 2, 2016. https://blog.paleohacks.com/cacao-vs-cocoa/#

Beverages

This chapter contains the "Eye Smoothie" recipe that, in my opinion, is the cornerstone in the foundation of rebuilding eye health. I will forever be grateful to Mark's naturopath for the practical advice that not only made it possible, but easy for him to consume sufficient kale and fish oil. He hates the taste of them alone, but he doesn't notice their flavor in the eye smoothie.

The naturopath's instructions for making the smoothies included (1) to put the blueberries into the blender solidly frozen, straight from the freezer (this releases maximum anthocyanin amounts from them) and (2) to chop the greens before adding them to the blender to release more of their nutrients. However, if you are a macular degeneration patient, you may not be able to safely chop large quantities of leafy vegetables. It is better for patients to let the blender do the chopping rather than risk injury or miss the high-level eye nutrition in the smoothies. The best thing for your eyes is to consume a daily eye smoothie. See page 88 for how to have the blender chop the vegetables.

The coconut milk in the smoothies contains fat which helps with the absorption of nutrients, but the naturopath said the fat in almond milk or other nut milk would also work. If you would like the smoothie to be a whole meal, she said to include almond butter or another natural nut butter to add protein to the smoothie. For those following the insulin resistance diet, add at least one but preferably two units of protein to each smoothie. Two units of protein is contained in four tablespoons of peanut or cashew butter or six tablespoons of almond butter. To use other nut and seed butters, read the nutrition facts on the label and add enough to contain 14 grams of protein.

The stevia or monk fruit and vanilla in the smoothie recipe were my ideas for making the smoothie more palatable. You may not need or want them, or you might substitute a different flavoring, possibly a citrus oil. (See "Sources," page 214, for where to purchase lemon, orange, or lime oil). If you would like to use vanilla but are allergic to yeast, and thus to the alcohol in most vanilla extract, use Frontier Naturals™ alcohol-free vanilla. See "Sources," page 214, for where to purchase this vanilla.

The smoothie can be used as a carrier for other supplements in addition to the fish oil. Aside from retaining enough greens, the blueberries, the fish oil and a fat-containing milk, feel free to modify it in any way that suits your taste. If you don't mind the taste of kale and plan to consume fruit high in anthocyanins at another meal, you can omit the blueberries for a zero-carb-unit smoothie. I know someone who often has his smoothies without the blueberries so he can eat toast with his breakfast and still consume only one carbohydrate unit in the meal.

There are also recipes for anti-inflammatory ginger beverages and thirst quenchers in this chapter. The "Sugar-free Cola" recipe on page 91 was very good for Mark's state of mind when he was adapting to changes in his diet. It contains no sugar, corn sweeteners, artificial sweeteners, caffeine, or other chemicals, unlike "the real thing." (Diet sodas are toxic to eyes and regular sodas will destabilize blood sugar levels. See page 60 and 44 to 45 for more about this). If yeast allergy prevents the use of the alcohol-based flavorings in this recipe, see a cola syrup recipe made from spices and the zest of oranges, lemons and limes here: https://cooking.nytimes.com/recipes/1013686-cola-syrup . Unfortunately, this recipe is labor intensive and I have not had time to try it yet, but I would suggest that it be made sugar-free by replacing the sugar and brown sugar with 2¼ teaspoons of stevia or pure monk fruit extract, or more or less to taste.

Go for it! Make the eye smoothie and take the most important step toward preserving or improving your vision.

Eye Smoothie

If you would you like days off from making a smoothie, use one of the larger batch ingredient lists and store the extra servings in the fridge for a few days.

Ingredients for 1 serving

   1 cup pre-washed baby spinach leaves (packed in metal measuring cup)
   2 large or 3 small-medium kale leaves
   ⅓ can (about 4½ ounces) light coconut milk OR ⅝ to ¾ cup almond or other nut milk
   1 teaspoon Nordic Naturals™ omega-3 fish oil (optional)[1]
   1 teaspoon vanilla (optional)
   3/16 teaspoon enzyme-treated stevia extract or pure monk fruit extract, or to taste (optional)
   ¾ metal measuring cup frozen blueberries
   Additional water or nut milk to produce desired thickness

Ingredients for 3 servings

   3 cups pre-washed baby spinach leaves (packed in metal measuring cup)
   5 large or 7 small-medium kale leaves (a bunch weighing about ½ pound)
   1 can (about 14 ounces) light coconut milk OR 1⅞ to 2¼ cups of almond or other nut milk
   1 tablespoon Nordic Naturals™ omega-3 fish oil (optional)[1]
   1 tablespoon vanilla (optional)
   ½ + 1/16 teaspoon enzyme-treated stevia extract or pure monk fruit extract, or to taste (optional)
   2¼ metal measuring cups frozen blueberries
   Additional water or nut milk to produce desired thickness

---

1 If fish oil is an allergy problem, try NutraVege™ Omega-3 Plant made by Nature's Way™.

Ingredients for 4 servings

> 4 cups pre-washed baby spinach leaves (packed in a metal measuring cup)
>
> 7 large or 10 small-medium kale leaves (a bunch weighing about ¾ pound)
>
> 1⅓ cans light coconut milk OR ½ can regular coconut milk plus ⅞ cup water OR 2½ to 3¼ cups any nut milk
>
> 1 tablespoon + 1 teaspoon Nordic Naturals™ omega-3 fish oil[1]
>
> 1 tablespoon + 1 teaspoon vanilla (optional)
>
> ¾ teaspoon enzyme-treated stevia extract or pure monk fruit extract, or to taste (optional)
>
> 3 metal measuring cups frozen blueberries
>
> Additional water or nut milk to produce desired thickness

Wash the kale, being sure to remove all dirt from the inner sides of the stems. Remove the large tough stems and chop the leaves into small pieces if you are "the cook." If you have low vision, tear the leaves, in about two inch pieces, from the thick, tough parts of the stem. Discard the stems. Place the kale pieces and spinach in the blender in small batches, filling the blender ¼ to ½ full (depending on how much your blender can handle and still chop almost all the leaves). Remove each batch from the blender to a light-colored bowl. When you finish chopping, put all the greens in the blender. Add the coconut milk, coconut milk plus water, or the nut milk to the blender and blend on high speed until the greens are pulverized. Add the fish oil, vanilla, and stevia or monk fruit and blend. Add the blueberries and blend thoroughly, stopping and scraping down the sides of the blender a few times. Add more water or nut milk if the smoothie is too thick. Divide the smoothies into the number of servings made and store servings you will not drink immediately in the fridge in glasses with covers. Makes as many servings as specified by the ingredient list. Each serving contains 1 carbohydrate unit. If you omit the blueberries, the smoothies contain zero carbohydrate units.

Anti-inflammatory Beverages

Ginger Tea by the Cup

Skip the ginger tea bags and make tea from fresh ginger root for the full benefit of its anti-inflammatory effects.

> 1 ounce fresh ginger root
> 1¼ cups water

Wash the ginger root and cut off any damaged or discolored spots. If you are rushed for time, dice it finely. If you have more time and want more potent tea, grate the ginger root. Combine the ginger and water in a saucepan and bring it to a boil. Reduce the heat and simmer it for 10 to 15 minutes. Strain out the ginger and enjoy the tea immediately. Makes one serving. This recipe may be doubled, tripled, etc. if it is for serving more than one person. To save work if drinking ginger tea often, make a large batch of ginger concentrate (recipe at the botom of this page) to make "Quick Ginger Tea," below. Contains zero carbohydrate units.

Quick Ginger Tea

> Ginger concentrate, below, liquid or made into frozen concentrate cubes
> Water

Fill a cup or mug about ⅙ to ¼ full with ginger concentrate or about ⅓ to ½ full of frozen ginger concentrate cubes. Add water to finish filling the cup or mug. Transfer the mixture to a saucepan to bring it to a boil or warm it in the microwave. Makes one serving which contains zero carbohydrate units.

Ginger Concentrate

With a food processor, you can easily make enough ginger concentrate to last for several weeks of ginger tea or ginger ale. Then you can enjoy these beverages easily when time is tight. Use the freshest ginger you can find – heavy and smooth skinned – for this recipe.

> 2 to 2¼ pounds fresh ginger root
> 10 cups water

Scrub the ginger root and cut off any damaged or discolored spots. Cut the ginger into chunks that will fit into the feed tube of a food processor. Grate it with the processor, cleaning the fiber out of the holes in the blade two or three times when you stop during processing. Reserve the fiber to cook with the grated ginger.

Combine the grated ginger with the water in a large stockpot and bring it to a boil over medium heat. Reduce the heat and simmer it for 20 to 30 minutes. Cool the mixture at room temperature until it is lukewarm or cooler. Working with about two cups of the cooked ginger mixture at a time, strain it through a wire mesh strainer, pressing the grated ginger with the back of a spoon to extract all of the liquid.

Freeze the concentrate in small jars containing enough concentrate for you to use in two or three days or put it into ice cube trays to freeze. When the ginger concentrate cubes are frozen, transfer them to a container and return them to the freezer. Makes 9 to 10 cups of concentrate, or enough for about 40 cups of tea. Contains zero carbohydrate units.

If this is more work than you prefer to do and you tolerate lemon juice, try Edwards & Sons™ organic ginger juice[2] instead of homemade concentrate. It is probably not as concentrated as this recipe, so taste test to determine how much to use in "Quick Ginger Tea," page 89, or "Ginger Ale," below.

Thirst Quenchers

The recipes below call for stevia or monk fruit for sweetening. My family prefers stevia in beverages because its sweet flavor is long-lasting. Taste perception of both stevia and monk fruit varies from person to person, so start with the lowest amount of sweetener in the recipe and add more until it tastes just right. See pages 82 to 83 for more about these sweeteners.

Ginger Ale

This refreshing traditional soda will cool you off on a hot summer day and is also anti-inflammatory.

> 2 to 4 tablespoons ginger concentrate on page 89, or to taste
> Choose one sweetener:
>> $1/16$ teaspoon enzyme-treated stevia extract or pure monk fruit extract, or to taste
>> ⅛ to ½ teaspoon stevia or monk fruit working solution, recipes on page 169
>> 2 to 3 teaspoons agave, or to taste (Use agave only occasionally such as when you have guests)
> 1 cup sparkling or carbonated water
> Ice

Measure 2 tablespoons of the ginger concentrate and the smaller amount of the sweetener into a glass and stir them together thoroughly. Add the carbonated water and stir. Taste the ginger ale and add more ginger concentrate or sweetener if desired. Record how much ginger concentrate and sweetener make it taste "just right" to you, perhaps in the margin of this page, for next time. Add ice and serve. Makes 1 serving.

---

2 To purchase the ginger juice, click here: https://store.edwardandsons.com/products/organic-ginger-juice for online sales or here: https://store.edwardandsons.com/pages/locator to find a store..

Lemonade or Fizzy Lemon Soda

Make lemonade early in the summer to determine the "just right" amount of sweetener for you. Then when summer heat comes, take a pitcher of this lemonade outside and relax.

⅔ cup lemon juice squeezed from two to three lemons

¼ to ½ teaspoon enzyme-treated stevia extract or pure monk fruit extract, or more to taste

3 to 4 cups cold water or chilled sparkling or carbonated water

Ice

Stir the juice and the smallest amount of sweetener together until the sweetener is completely dissolved. Add three cups of water, stir and taste. If it's too strong, add more of the water. If it's not sweet enough, add more sweetener and taste after each small addition. When you achieve "just right," record the amounts of water and sweetener used, possibly in the margin of this page, for next time. Serve with ice. Makes about 4 servings.

Sugar-free Cola

This sugar- and artificial sweetener-free cola tastes so good that you won't have any trouble kicking the soda habit. See "Sources," page 213 for where to get the special ingredients used. If you are allergic to yeast so can't have the alcohol in the vanilla and cola flavoring, see information on the next page for the New York Times recipe for cola syrup made from citrus fruit zests and spices.

Ingredients for 1 serving made in a 14 to 16 ounce glass

1½ teaspoons stevia or monk fruit working solution (or to taste in a range of 1 to 1¾ teaspoons), recipes on page 169

⅜ teaspoon cola flavor (See "Sources," page 213)

⅛ teaspoon vanilla, preferably Madagascar Bourbon vanilla. (See "Sources," page 213)

3-4 drops lemon juice

1 teaspoon caramel color working solution (optional, or to your preferred color), recipe on page 170

12 ounces of purchased sparkling water or home-carbonated water

If you are not using purchased carbonated water, carbonate the water with a SodaStream™. In a large glass, stir together the first five ingredients with about an ounce of water. Slowly pour the carbonated water into the glass by running it down the side of the glass to retain carbonation. Stir for just a second or two. Serve with ice if desired. Makes one serving which contain zero carbohydrate units.

Ingredients for 3 servings of about 12 ounces each or 1 liter of soda

> About 4½ teaspoons stevia or monk fruit working solution (ot a range of 3 to 5 teaspoons or to taste), recipes on page 169
>
> 1⅛ teaspoons cola flavor (See "Sources," page 213)
>
> ⅜ teaspoon vanilla, preferably Madagascar Bourbon vanilla, (See "Sources," page 213)
>
> ⅛ teaspoon lemon juice
>
> 2½ teaspoons to 1 tablespoon caramel color working solution (optional, or to your preferred color), recipe on page 170
>
> About 1 ounce of cold water to bring the volume up to ¼ cup
>
> About 1 liter of purchased sparkling water or home-carbonated water*

Before making a liter of soda, make a few singles glasses as on the previous page to determine how much sweetener you prefer. Multiply that amount by three for the amount to use in this 1-liter recipe.

If you are not using purchased carbonated water, carbonate the water. If you are using a SodaStream™, refrigerate a 1-liter bottle of purified water for a few hours or longer. Then carbonated it using the manufacturers directions. If you are using purchased carbonated water, remove about two ounces of carbonated water from the purchased bottle. (It will contain the same amount as a SodaStream™ bottle).

Measure the first five ingredients into a glass measuring cup. Add cold water to bring the amount in the cup up to ¼ cup and stir thoroughly. Tilt the bottle of carbonated water. Pour the liquid ingredients in the measuring cup into the bottle, pouring slowly down the inside wall of the bottle. Cap the bottle tightly and gently invert it once. Refrigerate until serving time or serve immediately with ice if desired. Makes 3 servings which contain zero carbohydrate units. This recipe may be doubled, tripled, etc. and the amount over ¼ cup refrigerated for making future 1-liter bottles of cola.

*Note about carbonated water: We tried several brands of sparkling water when I first made this soda, and Mark could taste bitterness (possibly from the minerals from the spring water) in all of them. Therefore, we purchased a SodaStream™ and used water purified by our filter to produce carbonated water. Then the cola tasted good to him. Carbonating water at home is more environmentally friendly and saves the effort of hauling bottles of water home from the grocery store.

The *New York Times* "Cola Syrup" recipe can be found here: https://cooking.nytimes.com/recipes/1013686-cola-syrup. I suggest substituting 2¼ teaspoons, or to taste, of stevia or monk fruit extract for the sugar and brown sugar. Perhaps add water to the final concentrate to bring the volume up to 12 ounces, and then use one ounce in each serving of cola.

Breakfasts

Mom was right – breakfast really is the most important meal of the day. A good protein-containing breakfast promotes stable blood sugar levels all day long, and stable blood sugar is essential for eye health. The recipes in this chapter are for traditional breakfast foods, but don't let tradition limit what you eat for breakfast. There is nothing wrong with a sandwich for breakfast. I eat game meat or "clean" turkey and vegetables for breakfast every day.

The first few recipes in this chapter are designed to be easy for macular degeneration patients to prepare for themselves. The last few are for special treats made when a family member or friend is the cook.

Get your day off to a good start with the good eye nutrition in these breakfast foods.

Walk-Away Oatmeal

Breakfast is ready any time you are if you start this cereal in a crock pot the previous evening. The cinnamon in this oatmeal promotes stable blood sugar levels as well as adding a bright flavor for a morning perk-up. Goji berries are loaded with nutrients for preserving and improving vision.

> 1 cup of thick rolled oats such as Bob's Red Mill™ extra-thick whole grain rolled oats (See "Sources," page 215. Also available processed in a gluten-free facility)
> 4 cups water (may prefer less for thick oatmeal if you skip the dried fruit)
> 1 cup raw pumpkin seeds or 1¼ cup nuts such as slivered almonds, shelled pistachios, or other small or chopped nut pieces of your choice
> No-sugar dried fruit such as ½ cup goji berries or raisins, cut up Turkish apricots, or other fruit of your choice cut into small pieces (optional)
> 1½ to 3 teaspoons cinnamon, or more to taste

Optional ingredients to serve with the oatmeal
> Milk or alternative milk
> Monk fruit or stevia extract powder or working solution, recipes on page 169

The evening before you wish to have oatmeal for breakfast, place all the ingredients in a 2 or 3 quart crock pot and stir to thoroughly combine them. Turn the crock pot on to the low setting. Walk away and let the crock pot cook for 8 to 9½ hours. Do not stir the cereal at all during the cooking time. In the morning at seving time, enjoy creamy oatmeal with milk if desired. If you added fruit, you probably will not need a sweetener. However, if you want more sweetness, use monk fruit or stevia extract powder sparingly (it is easy to add too much) or use the working solutions (easier to control) until the cereal tastes just right. Refrigerate leftover oatmeal and warm it in a saucepan or microwave it for future breakfasts.

Makes four servings, each of which contains 1 carbohydrate unit and 1 protein unit if the optional ingredients are omitted. This recipe may be doubled and still made in a 3 quart crock pot.

Walk-Away Hard Boiled Eggs

Eggs from naturally raised chickens have deep yellow yolks loaded with lutein for your eyes. You'll love their rich flavor also.

> Eggs, as many as you would like to eat in five or six days
> Water
> 1 teaspoon vinegar (optional, omit if allergic to yeast)

Place the eggs in the bottom of a saucepan in a single layer without crowding them. The only limit on how many eggs you can cook is the size of the pan. Add cool water to a depth of two inches above the tops of the eggs. If desired and you are not allergic to yeast, add the vinegar which may help prevent the eggs from cracking.

Place the pan on the burner of your stove and heat it on high heat until the water comes to a full rolling boil. Turn the heat off but leave the pan on the burner. Walk away for at least 12 to 15 minutes, although if you're gone longer, that is fine. When you come back, remove the pan from the stove, drain most of the water, place the pan in the sink and run cold water into it until the eggs are cool enough to handle. Remove the eggs that you would like to eat immediately and enjoy breakfast. Refrigerate the remaining eggs for up to five days. Each egg is one protein unit.

Breakfast Sausage Patties

Braising these patties in the oven keeps them tender no matter what kind of meat you use. Black pepper is anti-inflammatory, so add as much as you like for the sake of your eyes.

> 1 pound ground white-meat turkey, bison, grass-fed beef or other meat of
>   your choice
> 2 to 3 teaspoons fennel seeds
> ¾ teaspoon salt, or to taste
> ½ teaspoon pepper or more for inflammation, to taste
> Water

Mix all the ingredients together thoroughly using your hands. Shape the mixture into four to six patties. Place the patties in a single layer in a covered casserole dish and add water to a depth of ¼ to ½ inch. Cover the casserole with its lid and bake the patties at 350°F for 30 minutes. Uncover them and bake them an additional 15 minutes or until browned on top. Turn the patties over and bake them for another 15 minutes, or until the second side is browned on the top and the water has evaporated. With this cooking method, you may wish to make these patties the day before you plan to have them for breakfast and refrigerate them overnight. If a family member or friend is cooking and using a tender type of meat, these patties may be broiled 10 minutes on each side in the morning. Makes four servings of 3 protein units each or six servings of 2 protein units each.

Vegetable Frittata

This is delicious for dinner or as a high protein breakfast. If made with the pepper and tomato it contains carotenoids including lycopene.

Vegetables – Choose one
    ½ red pepper + ½ small tomato
    1 zucchini (about 6 ounces)
    1 cup frozen cauliflower florets, smaller ones if possible
    1 cup frozen broccoli pieces
    Leftover cooked asparagus or other vegetables

Egg mixture
    4 eggs, preferably from naturally raised chickens
    About 3-4 tablespoons of milk (optional)
    Dash of salt

For cooking
    About 1 tablespoon monounsaturated oil for the vegetables
    About 1 tablespoon butter or additional oil for cooking the eggs

The vegetables may be partially cooked before serving time and left in the pan a while or be partially cooked the evening before and refrigerated overnight. The directions for preparing each type of vegetables are:

Pepper-tomato mixture: Cut the pepper into ¾ inch squares and add them and the oil to the pan. Sauté on medium heat for 5 to 10 minutes, stirring occasionally. Turn the heat off, place a lid on the pan, and leave the pan on the warm burner for about 20 minutes until right before mealtime. Cut the tomato into ½ inch pieces and add it to the pan after the peppers have steamed long enough to be soft, and leave it until near meal time. Turn the burner back on a few minutes before adding the eggs and heat to the pan to begin cooking the tomato.

Zucchini: Slice the zucchini about ¼ inch thick and add the slices and oil to the pan. Sauté on medium heat for 5 to 10 minutes, turning the slices with a spatula. Let the zucchini sit in the pan and soften with the lid on until meal time.

Cauliflower or broccoli: Take 1 cup of cauliflower or broccoli out of the freezer the evening before or a few hours before the meal. Let the vegetables thaw in the refrigerator. If you forget to take them out, microwave briefly to thaw them. Put the vegetables in the pan with the oil and sauté for about 5 minutes. Leave then in the pan with the lid on until right before the meal. If you have leftover cooked cauliflower, broccoli, asparagus or other vegetable, they are ideal to use in this frittata.

If the vegetables were refrigerated overnight, drain any liquid from them. Add the butter to the pan with the vegetables, cover with the lid, and turn the heat on so the butter can melt and the vegetables and pan can warm while you prepare the eggs.

Beat the eggs, add the milk and salt, and beat again briefly.

When the butter bubbles, pour the eggs into the pan and cook for a few minutes until some egg is solidifying at the bottom of the pan. Use a spatula to lift the cooked egg and turn it over. Repeat this until the egg is all cooked. Makes two servings, each containing two units of protein and zero carbohydrate units.

Eggs Benedict

This is a very special way to get lutein from eggs. The hollandaise sauce on the next page is prepared in a microwave oven, making this gourmet breakfast or dinner relatively easy.

> 2 whole grain English muffins or 4 slices of whole grain bread you tolerate
> 4 one-ounce slices of Canadian bacon
> About 1 teaspoon oil or butter
> 4 large eggs, preferably from naturally raised chickens
> "Easy Hollandaise Sauce" on the next page

If you are using English muffins, split them in half. If you are using bread and you wish to decrease the amount of carbohydrate per serving, use a large (3½ inch) biscuit cutter to cut rounds out of the bread slices. Toast the muffins or bread. Place the Canadian bacon on a dish to microwave or in a skillet to heat it. Oil or butter the cups of an egg poaching pan or four silicone egg poachers. Measure the ingredients for the hollandaise sauce into a bowl, except for the lemon juice and salt.

Bring water to a boil in the bottom of the poaching pan or in a sauce pan if you are using silicone egg poachers. Warm the egg cups over the boiling water in the poacher. Microwave or warm the Canadian bacon. Break the eggs into the cups of the pan or the silicone egg poachers. Gently place the silicone egg poachers in the boiling water in the saucepan. Cover the poacher or saucepan with its lid. While the eggs are poaching, prepare the sauce as directed in the recipe on the next page.

While the sauce is microwaving, place the toasted muffins or bread on serving plates. Top each with one slice of Canadian bacon. When the eggs are cooked, loosen them from the cups with a spoon and place them on the Canadian bacon. Top with hollandaise sauce. Makes four servings each of which contains 1 carbohydrate unit or less[1] and 2 protein units or two servings containing 2 carbohydrate units (or less if the bread was trimmed) and 4 protein units.

---

1 Read the muffin or bread label to be sure the net carbohydrate per serving of the amount of bread used is under 15 grams. Net carbohydrate is the grams of total carbohydrates minus the grams of fiber.

Easy Hollandaise Sauce

If you have leftovers after using this sauce and are not planning to use it for Eggs Benedict soon, serve it on cooked broccoli or asparagus for a special treat.

2 teaspoons tapioca starch or arrowroot
Pinch of white pepper, or black pepper if white is not available (optional)
1 tablespoon water
1 cup plain yogurt
1 large egg, lightly beaten, preferably from naturally raised chickens
1½ tablespoons butter, cut into pieces
1 tablespoon lemon juice
Dash of salt

In a glass bowl, stir together the starch, pepper, and water. Add the yogurt and slightly beaten egg and whisk with a wire whisk. Add the pieces of butter. Put the bowl in the microwave oven and microwave on high power for three minutes, whisking well after each minute of cooking, or until the mixture is thickened. (Don't worry if the mixture looks curdled when you first remove it from the microwave before you have whisked it. It will become smooth as you whisk). Add the lemon juice and salt, whisk again, and microwave for an additional 30 seconds. Whisk and serve.

Makes about 1½ cups of hollandaise sauce which contains negligible carbohydrates, enough for about three recipes of "Eggs Benedict." Refrigerate the leftover sauce in a glass jar to serve with future meals. Re-warm leftover sauce in the microwave, beating it with a wire whisk to make it smooth after each 45 seconds of microwaving.

Main Dishes

Main dishes are the heart of our meals so should be high in the nutrients needed for eye health as well as being easy to prepare. The recipes in this chapter supply high-quality protein for healing, plus plenty of eye-supportive nutrients from the vegetables they contain. The recipes that use tomatoes also contain some fat to promote absorption of lycopene from the tomatoes. (See page 38 for more about lycopene). Most of the recipes in this chapter are either baked in the oven or prepared in a crock pot for ease and safety. The final recipe is a "comfort" recipe which takes more work but it cheers the heart of both the patient and the cook.

Conventional ovens are usually not considered to be work and time-saving appliances but they can be if you count *your* time rather than just how much time it takes the food to cook. Although the food will cook slowly (and have time for the flavors to blend deliciously), the amount of time you spend working on the meal can be brief. Some of the recipes in this chapter are entrées which can become part of an oven meal.

Oven meals usually consist of a main dish, a starchy side dish such as a grain or starchy vegetable, a non-starchy vegetable, and possibly a dessert all baked together in the oven at the same time. When I first tried them, I was amazed at how easy to make and how delicious oven grains and oven vegetables were – much tastier than their stove-top counterparts. All of these dishes, including the desserts, are more flexible than oven-baked foods such as bread: they can cook at a range of temperatures and for a range of times. Bake them for the same time and at the same temperature as the main dish and add a little more water if they will cook for a longer-than-usual time or at a higher temperature.

For oven meal recipes in addition to the oven entrées in this chapter, see pages 117 to 119 for oven vegetables, 122 to 125 for oven grains, and 150 to 151 for dessert recipes that work well with an oven meal.

Crock pots save both money and time. They use less expensive cuts of meat which will become flavorful and tender when they cook all day. Protein-rich, economical dried beans also produce easy meals prepared in a crock pot.

The recipes in this chapter were developed for a three-quart crock pot. If your crock pot is not that size, you can still use these recipes. If you are cooking for one or two people, reduce the recipe ingredients to half the given amount for two-quart crock pots.

However, rather than making small batches, I would advise investing about $25 to $35 for a five to six quart no-frills crock pot to make the best use of your time. If you cook a larger quantity of a crock pot entrée than you eat immediately, freeze the leftovers for future meals. On a busy day, happily use the leftover meal from the freezer. The best way to reduce your kitchen workload and simplify your life is to purchase a five to six-quart crock pot, double the recipes in this chapter, and freeze leftovers for future meals each time you use the crock pot. Avoid storing large amounts of frozen fruits and vegetables in your freezer, but rather use the space for pre-made meals and purchase frozen grocery store items week-by-week.

Beans are excellent sources of protein, fiber, vitamins and minerals and take little of your time to prepare. However, they do require some advance preparation. The day or evening before you will serve the recipe, begin soaking the beans overnight or longer. It is also possible to prepare fresh vegetables and other ingredients ahead if more convenient. In the morning, rinse the beans, add the other ingredients to the pot, turn it on, and a delicious, wonderful smelling dinner will be ready to eat at the day's end.

Enjoy these main dishes and stocking your freezer with meals for future nearly-effortless dinners.

Easy Baked Salmon

*This recipe tells how to cook salmon without the visual demands of broiling and how to easily make it skinless without paying to have the skin removed. For omega-3 fats, eat wild salmon only because **farmed salmon is high in pro-inflammatory omega-6 fats.** (Read more about this on pages 38 to 39). For variety, try the "Sweet Salmon" variations on the next page.*

> 1 pound of wild-caught salmon fillet
> 2 to 4 teaspoons butter, olive oil, or other monounsaturated oil
> Salt and/or pepper (optional)

Preheat your oven to 400°F. Rub the bottom of a glass baking dish with about ½ teaspoon of the butter or oil. Lay the salmon in the baking dish skin side down. Bake for about 8 minutes or a little longer for very thick fish. When the baking time has finished, remove the baking dish from the oven and place it on a heat-safe surface. Turn the fish with a spatula and, if desired, remove the skin[1] with a knife and/or your fingers. Dot the fish with more of the butter or drizzle it with the oil. Place the baking dish back in the oven and bake until the fish is opaque throughout and flakes easily with a fork, an additional 10 to 20 minutes. If you have difficulty seeing if the fish looks opaque, use a talking thermometer to tell you when the internal temperature reaches 145°F. It is better to err on the side of cooking the fish too long (to a higher temperature) than to eat it underdone, which can result in illness. Sprinkle the fish with salt and pepper if desired. Serve immediately. Makes 3 to 4 servings. Each serving contains the same number of protein units as the weight of the portion of cooked fish in ounces.

---

1 To derive the maximum amount of omega-3 fatty acids from salmon, leave the skin on so the oils under the skin soak into the flesh during cooking. Mark ate salmon broiled with the skin on for the first six months after his diagnosis. Then, as might be expected if you *have to* eat something often, he dreaded eating salmon and wanted it skinless. I had it skinned at the fish counter only once. When prepared using this recipe, he gets plenty of astaxanthin from the salmon and some of the omega-3 fats in a way he will eat, which is better than not eating it.

Sweet Salmon Variations of Easy Baked Salmon

Additional ingredients – chose one option:

Orange flavored:
    1 teaspoon orange juice concentrate
    1 teaspoon apple juice concentrate.

Lemon flavored:
    1½ teaspoons lemon juice
    ¾ teaspoon agave

Reserve 2 to 3 teaspoons of the butter or oil called for on the previous page and melt the butter. Stir the orange and apple juice concentrates or the lemon juice and agave into the oil or melted butter thoroughly. After removing the skin, drizzle the fish with the mixture. Finish cooking the fish as directed on the previous page.

Salmon Burgers

Relieve the boredom of "plain salmon" by making it into burgers and serving them on whole grain buns with lettuce, sliced tomato, pickles and condiments. (See allergy-friendly condiment recipes on pages 168 to 169 and buns on page 147). Always purchase wild-caught salmon for the maximum anti-inflammatory effect from its omega-3 fats and the natural astaxanthin wild caught salmon contains.

Single burger recipe which may be doubled, tripled, etc.:
    1 6-ounce can pink wild-caught salmon such as such as Wild Planet[TM2]
    ½ to 1 tablespoon arrowroot or tapioca starch (optional)
    1 tablespoon oil
    Dash to ⅛ teaspoon pepper, or to taste (optional)
    ⅛ teaspoon salt, or to taste (needed if using no-salt salmon)

Four-burger economy recipe:
    2 14.75-ounce cans pink wild-caught salmon such as Bumble Bee[TM]
    2 to 3 tablespoons arrowroot or tapioca starch (optional)
    3 to 4 tablespoons oil
    ¼ to ½ teaspoon pepper, or to taste (optional)
    ½ teaspoon salt, or to taste (needed if using no-salt salmon)

Drain the liquid from the salmon and remove any skin and vertebral bones from the meat. The small rib bones will crush and break down during cooking and are a

---

2 Wild Planet[TM] salmon is boneless and skinless.

good source of calcium, so leave them in the fish. Add the rest of the ingredients and mix thoroughly with your hands. If you use the starch, the burger will stay together better.

Oil the bottom of a glass casserole dish with a little additional oil. Form the mixture into one burger (if using the top set of ingredients) or four burgers (for the economy recipe) and place them in the casserole dish in a single layer. Bake at 400°F for 35 to 45 minutes or until they form a light brown crust on the bottom. Turn them and bake another 15 minutes until there is a crust on the other side. Serve alone or in a bun with condiments. Each burger contains 4 to 5 protein units or the same number of units as the ounces of weight of the cooked burger. The starch adds about 4 to 5 grams of carbohydrate to the burger, or $1/3$ of a carbohydrate unit.

High Nutrient Meatloaf

The red pepper and carrots in this meatloaf provide carotenoids for eye health and also give it such an excellent flavor that it doesn't really need the optional ketchup. However, if you can have tomatoes, the ketchup adds lycopene.

> 1 pound ground meat of any kind – grass-fed lean beef, bison, turkey,
>   venison or other game meat
> 1 to 2 slices onion, finely diced (optional)
> ½ red bell pepper, finely diced
> 1 cup grated carrots
> ¼ cup arrowroot or tapioca starch (may omit to reduce carbohydrates if
>   needed)
> ¾ teaspoon salt, or to taste
> ¼ teaspoon pepper (optional)
> ¼ teaspoon dry mustard (optional)
> ¼ cup water or ½ cup water, divided
> ¼ cup ketchup, (optional – see "Homemade Ketchup," page 168)

Place the ground meat, grated or diced vegetables and seasonings in a large bowl or 2- to 3-quart casserole dish. If you do not need to strictly limit carbohydrates and want a more cohesive meat loaf, also add the arrowroot or tapioca flour and ¼ cup of water. Thoroughly mix all the ingredients together with your hands and shape the mixture into an oval loaf. Place the loaf in the center of the 2- to 3-quart covered casserole dish and add ¼ cup water. Cover the casserole dish with its lid and bake the meatloaf at 350°F for 45 minutes. Then uncover it and bake it for another 30 minutes. Top it with the catsup during the last 15 minutes of baking, if desired. Makes 5 servings, each containing 2 protein units and ½ carbohydrate unit if the starch is used.

Quinoa Stuffed Peppers

Because quinoa contains high-quality protein, this is a satisfying vegetarian main dish. It also provides carotenoids from the peppers and lutein from the spinach or arugula. You can substitute three cups of any other cooked whole grain for the quinoa and water if you wish. Round out a vegetarian meal with some legumes (crock pot baked beans, page 116, are ideal) for additional protein to balance the carbohydrate in the quinoa.

     1½ cups quinoa
     3 cups water
     1 pound chopped spinach or arugula OR thawed and drained frozen spinach
     2 tablespoons oil
     7 red or orange bell peppers
     1½ teaspoons salt
     ¾ teaspoon pepper
     3 teaspoons sweet basil
     2 tablespoons paprika (optional - for color)
     Additional oil

Wash the quinoa thoroughly by putting it in a strainer and rinsing it under running water until the water is no longer foamy. Combine it with 3 cups of water in a saucepan. Bring it to a boil, reduce the heat, and simmer it for 15 to 20 minutes. Wash the fresh spinach or arugula and cut it into pieces. (The frozen spinach will not need chopping). Place the 2 tablespoons of oil in a saucepan, adding no water. Add the chopped greens and cook, stirring constantly, for a few minutes, or until the vegetables are wilted. While the vegetables and quinoa are cooking, use a paring knife to core the peppers, removing the stem, core, and seeds. Mix the quinoa and spinach with the seasonings and stuff the mixture into the peppers.

Lightly rub the bottom and sides of a casserole dish with the additional oil. Stand the peppers upright, with the stem hole at the top, in the casserole dish. Place the lid on the casserole dish and bake the peppers at 350°F for 45 to 55 minutes. Makes 7 servings, each of which contains 2 carbohydrate units and 1 protein unit.

Legume Soups

The soup recipes below and "Crock Pot Baked Beans," page 116, contain tomatoes which provide lycopene that is best absorbed with fat. Therefore, the recipes specify adding some oil to the crock pot near the end of the cooking time. However, if a salad with oil-and-vinegar dressing or crackers with butter are consumed at the same meal, the oil does not need to be added to the soup. I personally like to add a little oil to my bowl and stir it into the soup for richness and staying power.

These soups take very little of your time to make – possibly 5 minutes to get the beans soaking the evening before you want to serve the soup, 15 to 20 minutes the next morning to rinse the beans, prepare vegetables and fill the pot, and 5 minutes near serving time. Be sure to rinse the beans after soaking to remove indigestible carbohydrates that can cause intestinal gas. For further time-saving, make a double batch and freeze the leftovers.

Multi-Bean Soup

This soup is delicious and economical plus contains carrots and tomatoes for vitamin A, carotenoids and lycopene.

> 1 pound mixed dry beans*
> Water
> 5 carrots
> 3 stalks of celery
> ¼ teaspoon pepper (optional)
> 1 15-ounce can diced tomatoes or 1 11-ounce can tomato puree
> 2 tablespoons of monounsaturated oil such as olive, avocado or canola oil
>   (Optional - to improve absorption of lycopene, bottom of page 102)
> 1½ to 2 teaspoons salt, to taste

The evening before you plan to serve this soup for dinner, put the beans in a strainer and run water over them thoroughly to remove any dirt. Put them in a 3-quart crock pot and fill the pot nearly full with water. Soak them overnight. If you will be pressed for time in the morning, prepare the carrots and celery the evening before as described below and refrigerate them overnight.

In the morning, drain the water from the beans in the crock pot. Refill the pot with water and drain it two times more.* Add 5 to 6 cups of water to the pot and turn it to high. Peel the carrots and slice them into ¼ inch slices. Wash the celery and slice it into ½ inch pieces. Add the carrots and celery to the crock pot. Cook the soup on high for 5 to 6 hours or, if you are leaving for the day, turn it down and cook it on low for 8 to 10 hours.

When the beans are tender (after 5 to 6 hours if cooked on high), or slightly before serving time (if cooked on low), add the oil, salt, pepper, and tomatoes. If you are not using the diced tomatoes and the soup seems too thick, add up to 1 cup of boiling water. If you are adding the tomatoes or tomato puree shortly before dinner time, warm them in the microwave before adding them to the pot to speed the reheating time. Cook the soup on high for ½ hour or until it is heated through.

This recipe makes about 2½ quarts of soup, 4 servings which each contain 3 protein units or 6 servings which each contain 2 protein units. The carbohydrates in cooked legumes are mostly indigestible or fiber so this recipe contains zero carbohydrate units.

If you want to make a larger batch to freeze, use a 5-quart crock pot, double the amounts of all of the ingredients, and cook the soup on high for 8 to 10 hours. A double batch makes about 5 quarts of soup, or 8 to 12 servings. Leftovers freeze well.

*Notes about beans: Be sure that the bean mix does not contain barley or other gluten-containing grains or grains to which you are allergic. Bob's Red Mill™ 13-bean mix is all beans. Rinsing the beans after soaking them removes indigestible carbohydrates that can cause intestinal gas.

Crock Pot Lentil Soup

The carrots and tomatoes in this soup provide vitamin A and carotenoids including lycopene. Add cooked grains to this soup for variety.

    1 pound dry lentils
    Water
    3 to 5 carrots
    3 stalks celery
    ¼ teaspoon pepper (optional)
    1 15-ounce can diced tomatoes (optional but good for lycopene)
    2 tablespoons of monounsaturated oil such as olive, avocado or canola oil
        (Optional - to improve absorption of lycopene, bottom of page 102)
    2 teaspoons salt

The evening before you plan to serve this soup for dinner, rinse the lentils by running water over them in a strainer to remove any dirt. Put them in a 3-quart crock pot and cover them with cold water. Soak them overnight. If you will be pressed for time in the morning, prepare the carrots and celery the evening before as described below and refrigerate them overnight.

In the morning, drain the water from the lentils, add fresh water to the pot, and drain the water again. Rinse the lentils this way three times to remove indigestible carbohydrates which can cause gas. Drain off all the water after the last rinse. Peel the carrots and cut them into ¼ inch slices. Slice the celery into ¼ inch slices. Add 5 cups of water to the crock pot plus the carrots, celery and pepper.

Cover the crock pot and cook the soup on high for 6 hours or on low for 8 to 10 hours. Add the salt, oil and tomatoes an hour before the end of the cooking time. Check the soup near the end of the cooking time and add boiling water to the pot if the soup is thicker than you prefer. Makes about 2½ quarts of soup, 6 servings which each contain 3 protein units or 9 servings which each contain 2 protein units. The carbohydrates in cooked legumes are mostly indigestible or fiber so this recipe contains zero carbohydrate units. Leftover soup freezes well.

Black Bean Soup

This zesty soup contains anti-inflammatory black or chili pepper plus toma-toes for lycopene.

> 1 pound dry black beans
> Water
> 2 bell peppers, preferably one red and one orange, seeded and cut into
>   small pieces
> 1 small onion, diced (optional)
> 1 teaspoon ground cumin (optional)
> ½ teaspoon black pepper or a 2-inch chili pepper or chile pequin, seeded
>   and crumbled
> 2 teaspoons oregano
> 1 15-ounce can diced tomatoes or 1 pound tomatoes, chopped
> 2 tablespoons of monounsaturated oil such as olive, avocado or canola oil
>   (Optional - to improve absorption of lycopene, bottom of page 102)
> 2 teaspoons salt

The night before you plan to serve this soup for dinner, rinse the beans by running water over them in a strainer to remove any dirt. Put them in a 3-quart crock pot and cover them with cold water. Soak them overnight. If you will be pressed for time in the morning, chop the peppers and onion the evening before as described below and refrigerate them overnight.

In the morning, drain the water from the beans, add fresh water to the pot, and drain the water again. Rinse the beans this way three times. Drain off all the water after the last rinse. Add 4 cups of water to the crock pot plus the bell peppers, onion, cumin, pepper, and oregano. Cover the crock pot with its lid and cook the soup on high for 6 hours or on low for 8 to 10 hours. Add the oil, salt and diced or chopped tomatoes an hour before the end of the cooking time, or microwave the tomatoes to warm them and add them a few minutes before serving time. If the soup is thicker than you prefer, add a little boiling water to the pot near the end of the cooking time.

This recipe makes about 2½ quarts of soup, 5 servings which each contain 3 protein units or 8 servings which each contain 2 protein units. The carbohydrates in cooked legumes are mostly indigestible or fiber so this recipe contains zero carbohydrate units. Leftover soup freezes well.

Colorful White Bean Soup

This recipe is a take-off on a white bean and escarole soup I developed over 20 years ago. The green cabbage and escarole in the original recipe have been replaced with more colorful vegetables for their lutein, zeaxanthin, and carotenoids.

1 pound small white beans such as navy beans
½ small head of purple cabbage, chopped (about ¾ pound)
1 tablespoon fresh chopped sweet basil or 1 teaspoon dry sweet basil
¼ teaspoon pepper (optional)
1 small bunch of dandelion or chicory greens, washed and cut into 2-inch
    long pieces (about ¾ pound)
1 large white sweet potato (about 12 to 13 ounces), peeled and cubed
2 teaspoons salt

The day or evening before you plan to serve this soup for dinner, rinse the beans by running water over them in a strainer to remove any dirt. Put them in a 3-quart crock pot and cover them with cold water. Soak them overnight. If you will be pressed for time in the morning, prepare the vegetables as directed below the evening before. Refrigerate the cabbage and greens *separately in two containers* overnight. Put the cubes of sweet potatoes in water and refrigerate.

In the morning, drain the water from the beans, add fresh water to the pot, and drain and rinse the beans this way three times. Add the cabbage, basil, pepper, and 5½ cups of water to the crock pot. Put the lid on the crock pot and cook the soup on high for 4 hours or on low for 6 to 8 hours or until the beans are tender. Peel and cube the potatoes, wash and cut the dandelion greens. Add the potatoes to the pot about 2 to 2½ hours before mealtime. Add the greens about 1 to 1½ hours before mealtime. Then cook the soup another hour or so until the potatoes are tender. Add the salt at this point, shortly before mealtime. Check the soup and add a little boiling water to the pot if the soup is thicker than you prefer. Cook for the final half hour to allow flavors to blend. Makes about 2½ quarts of soup or 8 servings which each contain 2 protein units. The carbohydrates in cooked legumes are mostly indigestible or fiber so this recipe contains zero carbohydrate units.

Colorful Lentil Soup variation: Replace the white beans with green lentils, the cabbage with 1½ cups of chopped fresh tomatoes or a 14-ounce can of diced tomatoes, and the greens with ¾ pound of fresh baby spinach, except delay adding the tomatoes until about 1½ hours before mealtime. To improve absorption of lycopene from the tomatoes, add 2 tablespoons of monounsaturated oil when you add the salt.

Chana Dal Vegetable Soup

Chana dal is a variety of garbanzo beans that has an extremely low glycemic index score. Some diabetics swear by it for keeping their blood sugar on an even keel.

1 pound chana dal, such as Bob's Red Mill™ brand
Water
1 pound carrots
4 stalks celery
½ cup chopped onion (optional or to taste)
½ teaspoon pepper (optional or to taste)
1½ teaspoons sweet basil (optional)
1 15-ounce can diced or crushed tomatoes
2 tablespoons of monounsaturated oil such as olive, avocado or canola oil
    (Optional - to improve absorption of lycopene, bottom of page 102)
1½ teaspoons salt, or to taste

Pick over the chana dal beans and rinse them under running water in a strainer. Soak them overnight in at least four times their volume of water in a 3-quart crock pot. If you will be pressed for time in the morning, prepare the vegetables as described below and refrigerate them overnight.

In the morning, drain the soaking water from the beans. Replace it with cool water and drain and rinse the beans three times. Add 5 cups of water to the drained beans in the crock pot. Peel and slice the carrots. Slice the celery and chop the onion. Add them to the pot with the pepper and basil. (Do not add the oil, salt and tomatoes until near the end of the cooking time). Cook the soup for 6 to 8 hours on high or for 8 to 10 hours on the low setting. Mash a few of the beans against the side of the pot to thicken the soup. (Do not cook the soup for an excessively long time after mashing the beans). Add the oil, tomatoes and salt. If the soup seems too thick for you, add boiling water to bring it to the thickness you prefer. Cook on "high" for ½ hour or until it the soup is re-heated. (To speed the re-heating time, heat the tomatoes before adding them to the pot rather than adding them at room temperature).

Makes about 2½ quarts of soup, 4½ servings which each contain 3 protein units or 7 servings which each contain 2 protein units. The carbohydrates in cooked legumes are mostly indigestible or fiber so this recipe contains zero carbohydrate units.

Crock Pot Chili

This recipe can be mild or hot. My original version is mild. My son Joel makes his chili hot by substituting salsa for the tomato sauce and adding jalapeno peppers and crumbled dry chile pequin peppers for heat and flavor; his variations on the recipe are included.

> 1 pound dry kidney beans
> Water
> 1½ to 2 pounds of lean ground turkey, lean ground beef, buffalo,
>   or game meat
> 1 12-ounce can tomato paste
> 1 16-ounce can tomato sauce (For hot chili, substitute 2 cups of salsa).
> 1 small onion, chopped (optional)
> 1 teaspoon salt
> 1 to 3 teaspoons of chili powder, to taste, or 2 to 5 crumbled 1-inch long
>   dried chile pequin peppers, to taste (Use the chile pequin for hot chili).
> 2 to 4 jalapeno peppers, chopped (optional – omit for mild chili and use
>   with the seeds for the hottest chili).

The evening before you plan to serve the chili, rinse the beans by running water over them in a strainer to remove any dirt. Put them in a 3-quart crock pot and cover them with cold water. Allow the beans to soak overnight.

Preparation of the meat should be done the evening before if you will not be home at least an hour or two before dinner, or it can be done the afternoon of the day you serve this recipe if time allows. In a separate pan on the stove, brown the ground meat. Add the optional fresh onion and jalapeno peppers and cook for a few minutes more. Drain and discard the fat. If you are doing this the evening before, refrigerate the meat mixture.

In the morning, drain the water from the beans, add fresh water to the pot, and drain the water again. Rinse the beans this way three times. (This soaking and rinsing process removes difficult-to-digest carbohydrates which can cause intestinal gas). Add enough water to the beans in the pot to cover them. Cook on high for 4 to 6 hours or on low for 8 hours, or until the beans are tender.

Cook the meat and optional vegetables as above if not done the evening before.

When the beans are tender, drain the water until it is about half the level of the beans in the pot. Spoon tablespoons of the tomato paste into the pot. Add the cooked meat or meat and vegetable mixture, tomato sauce or salsa, and seasonings. Stir all of the ingedients together. If you like your chili juicer, add some boiling water.

If you are adding these ingredients shortly before dinner time, cook the chili on high for 30 to 60 minutes or until heated through. Warming refrigerated meat in the microwave can speed the heating time.

If you add these ingredients when mealtime is hours off, cook the chili for another 1 to 2 hours on high and stir the pot one or two times, or turn it down to low once it is thoroughly warm.

Makes 8 to servings, each of which contains 4 protein units. The carbohydrates in cooked legumes are mostly indigestible or fiber so do not add carbohydrate units.

If you wish to make a larger batch of this chili in a 5-quart crock pot, multiply the ingredient amounts by 1½.

Golden Stew

This satisfying stew is easy to make and contains plenty of good protein, vitamin A and carotenoids including lycopene.

> 2 to 3 pounds of beef, elk, antelope, venison, buffalo, or other game meat round steak, cut into cubes
> 1½ pounds carrots
> 1 pound orange or white sweet potatoes
> ½ cup quick-cooking (graulated) tapioca
> 1½ to 2 teaspoons salt, possibly divided
> ¼ teaspoon pepper (optional)
> 2 whole cloves (optional)
> 1 bay leaf (optional)
> 1 cup water
> 1½ pounds butternut squash, pared, seeded, and cut into 1 inch cubes (optional)

If you are using game meat a relative or friend hunted, the day before you plan to serve this stew, rub the frozen game meat with 1 teaspoon of the salt and allow it to thaw overnight in the refrigerator.[3] In the morning, peel the carrots and cut them into one-inch slices. Peel the sweet potatoes and cut them into cubes. Cut the meat into cubes, trimming off the fat, and combine it with the vegetables, tapioca, bay leaf, salt (or the remaining 1 teaspoon with wild game meat), pepper and water in a 3-quart crock pot. Stir the mixture well to evenly distribute the tapioca and cook it on low for 8 to 10 hours or on high for 6 hours. If you are using the squash, peel, seed and cube it and add it to the pot about 2 hours before the end of the cooking time. Any leftover stew freezes well. Makes 8 servings, each of which contain 2 carbohydrate units and 4 to 6 protein units. (Protein units vary with the amount of meat used).

---

3 Thawing with salt makes hunted wild game taste less gamey. Removing all the fat also helps gaminess. Beef, buffalo (bison) and farmed game meat do not need these treatments.

Crock Pot Roast Dinner

This recipe makes inexpensive, less-tender cuts of beef taste like gourmet fare. It also supplies good protein and carotenoids including lycopene.

> 1 2- to 3-pound chuck roast, rump roast, or pot roast of beef, buffalo (bison) or other red game meat
> 1½ to 2 pounds sweet potatoes, peeled and cut into cubes, OR small red potatoes
> 6 carrots or a combination of 3 to 4 carrots and 3 to 4 parsnips, peeled and cut into 2-inch pieces
> 1 onion, peeled and sliced or cut into eighths (optional)
> ½ cup water, beef bone broth, or red wine
> 2 tomatoes, chopped, or 2 tablespoons tomato paste or sugar-free ketchup (See the ketchup recipe on page 168 or purchase at a health food store).
> Dash of salt
> Dash of pepper (optional)

Peel and cut up the vegetables and put them into the bottom of a three quart crock pot. Set the roast on top of the vegetables. Stir together the water, broth, or wine with the tomato, tomato paste or ketchup. Pour the liquid over the roast. Sprinkle the roast with the salt and pepper. Cook on low for 10 to 12 hours or on high for 5 to 8 hours. This recipe is very tasty when made with a red wine such as Marsala. The alcohol evaporates during cooking leaving just its flavor, but if you are allergic to yeast, omit the wine. If you prefer more juice with your roast, increase the amount of water, broth or wine to 1 cup and double the amount of tomato, tomato paste or ketchup. Makes 6 servings, each of which contains 2 carbohydrate units and 5 to 7 protein units. (Protein units vary with the amount of meat used).

High Protein Lasagne

Although this recipe is more work than some, it provides a lot of cooked tomatoes for much lycopene and is a special treat that lifts the mood.

Sauce ingredients:

> 2 pounds lean ground turkey, beef, buffalo, or other meat
> 2 12-ounce cans tomato paste
> 1 28-ounce can tomato puree
> 1½ cups water
> 1 teaspoon salt
> ⅛ teaspoon pepper (optional)

Additional ingredients:

About 2 to 2½ boxes no-boil lasagne noodles such as Barilla™, or any gluten free or allergy-friendly no-boil pasta you tolerate

3 15-ounce containers of part-skim ricotta cheese, any type you tolerate (goat, soy, cow's milk etc.)

½ cup grated Romano cheese, optional (Use all sheep Romano if you are allergic to cow's milk).

32 ounces of part-skim mozzarella cheese or alternative cheese such as almond cheese, Funny Farm™ goat mozarella, etc.

1½ tablespoons chopped dried parsley (optional)

Start making the sauce early in the morning on a day when you have adequate time to cook if you want to serve, refrigerate, or freeze the lasagne that evening. Or, if you want to spread the work out a little, make the sauce one to a few days before you plan to serve, refrigerate, or freeze this lasagne.

Combine the tomato paste, tomato puree, water, salt and pepper in a 3-quart crock pot and turn it to "high." Cook the meat in a frying pan over medium heat, breaking it up and stirring it often, until it is well browned. Pour off and discard the fat. Add the meat to the crock pot and stir thoroughly. If you are having the lasagne the next day or later, you may turn the temperature control on the crock pot to low and allow the sauce to cook unattended for 8 to 10 hours total cooking time. However, if you want to shorten the cooking time and/or have thicker sauce, you may cook it on high for four to six hours, stirring it every 1 to 1½ hours. Cooking the sauce on high will allow you to assemble the lasagne the same day that you make the sauce and serve it for dinner.

When the sauce is almost cooked, I do not stir it within about the last hour of the cooking time. This allows the meat to settle so the sauce at the top of the crock pot is lower in meat content than the sauce at the bottom. I use a large spoon to remove some of the low-meat sauce from the top of the pot to use both under the first layer of pasta and on top of the last layer of pasta when I assemble the lasagne. After I've taken off enough of the lower-meat sauce, I stir the remaining sauce before removing it from the pot. However, if you like meaty sauce at the top and the bottom of your lasagne, stir it thoroughly before taking any from the pot. Homemade spaghetti sauce is also good for the top and bottom of the lasagne.

This size batch of lasagne sauce plus the cheese filling contains 75 protein units. If you use it to make 18 servings of lasagne, each serving contains 4 protein units. See the last paragraph of this recipe to determine how many servings you should divide the lasagne into with the type of pasta you use to get the correct number of carbohydrate units for glycemic control if needed.

The sauce may be refrigerated overnight or frozen. Makes about 2½ quarts of sauce. If you wish to make lasagne often, make a double batch of sauce in a 5-quart crock pot and freeze some of the sauce. Then when you want to make more lasagne, the sauce will be ready to use.

Lasagne assembly instructions:

In a bowl, combine the ricotta cheese, Romano cheese, and parsley. Mash them together with a potato masher. Slice the mozzarella cheese thinly.

Spread about 1½ to 2 cups of the sauce (total amount for all pans) over the bottoms of a deep 13 inch by 9 inch cake pan or a similar-sized deep casserole dish PLUS a 2½ to 3-quart casserole dish. If you have a large rectangular stainless steel lasagne pan/roaster which will hold the whole batch, spread the 1½ to 2 cups of the sauce in the bottom of it. Place the dry lasagne pasta in a single layer over the sauce. (The amount of pasta you need may vary with the type of baking dishes you are using). Spread the pasta with about 3 to 3½ cups of the sauce (total amount for all the pans). Layer about half of the mozzarella cheese over the sauce. Spread half of the ricotta mixture over the mozzarella. Add another layer of pasta to the dish(es), followed by another 3 to 3½ cups of sauce, the rest of the mozzarella, and the rest of the ricotta mixture. Add a third layer of pasta to the dish(es) and top it with about 2 cups of sauce. If you wish to, you may refrigerate or freeze the lasagne at this point. Cover the dish(es) with plastic wrap to refrigerate or freeze the lasagne

If frozen, before baking the lasagne, thaw it completely in the refrigerator. Cover it with aluminum foil or with the lid of the casserole dish. Bake the lasagne at 350°F for about 1½ hours or until it is hot throughout and bubbly at the edges. If you refrigerated or froze the lasagne before baking it, allow about an extra ½ hour or more of baking time. If it was frozen and is not completely thawed before baking it, add even more baking time.

This recipe makes a large batch for holiday dinners or to stock your freezer for future meals. If freezer space is limited, make only part of the sauce into lasagne and freeze the leftover sauce. Thaw the sauce and assemble additional pans of lasagne quickly at a later date.

The amount of this lasagne that is 2 carbohydrate units varies depending on the brand of pasta used. For Barilla no-boil lasagne pasta, if you cut the lasagne into squares that are half the length of a piece of pasta, 1½ of these squares is a little less than 2 carbohydrate units. Because this dish contains a generous amount of meat and cheese, there will be enough protein for balance in each serving (4 units of protein if you make 18 servings). The full batch of lasagne sauce and the cheese fillings contains 75 protein units.

Vegetables

Vegetables are nutritional powerhouses. They are great sources of vitamins, minerals, carotenoids, phytonutrients, and pigments needed by eyes such as lutein and zeaxanthin. They also contain fiber for good digestion. Some, such as dried beans, even contain a good quantity of protein. People who eat five servings of vegetables or fruits per day have a 30% less risk of having a stroke than people who do not. Obviously, vegetables are foods you will want to plentifully consume for good general and eye health. This chapter provides basic techniques for preparing the vegetables most important for your eyes. It also includes recipes for vegetable treats.

Braised Greens

Use this recipe to prepare any type of greens, from the standard eye-doctor recommended kale, collards and spinach to Swiss chard to less common types of greens such as chicory or dandelion. Many of these greens are delicious eaten raw, especially when the leaves are young and small.

> 1 bunch of greens of any kind including, but not limited to, those above
> 2 teaspoons monounsaturated oil such as olive, avocado, or canola

Many greens have stems that will soften when cooked. However, if your kale or other mature greens have extremely tough or woody stems, remove them. Cut or tear large leaves of greens into smaller pieces of about 3 inches. Put the leaves in a sink filled with cool water. Add one or more teaspoons of salt[1] to the water especially if the greens are home-grown or from the farmers' market. Swish the greens in the water, lift them out, and allow the water to drain from each handful. Place them in the other side of the sink. Fill the second side of the sink full of water, and repeat this three or more times until there is no more dirt or grit left in the sink.

Heat the oil in a large saucepan. Put the leaves into the pan with water still clinging to the leaves. Cook the greens on medium heat, stirring them often so they all spend some time at the bottom of the pan, until all the greens have wilted. If the stems are not tender, cover the pan with its lid, turn off the stove, and leave the pan on the burner for 10 minutes or longer. Some greens, such as chicory or dandelion greens, lose much of their bitterness when cooked this way.

Swiss chard, with its large stems, becomes a real treat when given extra attention. After washing the chard, cut the stems out of the leaves and slice them into one-half to one inch pieces. Add only the sliced stems to the pan with the oil and

---

1The salt is added to immobilize any insects that might be in the greens so they relax their hold on the leaves and can be washed away. When I was a child and washed greens, a sprinkle of salt from a large shaker was enough. Now I add at least one teaspoon or more during earwig season. This may be a testimony to the toughness of earwigs, or it might be a result of all the pesticides insects have been exposed to in the last half of the 20th and the 21st centuries.

sauté the stems until tender. If all of the oil has been absorbed, add more oil. Then add the leaves, cut into pieces, to the pan and cook them as above.

To derive the greatest amount of nutrients, do not salt dandelion greens while cooking them. According to the most helpful table of lutein-zeaxanthin values (http://foodinfo.us/SourcesUnabridged.aspx?Nutr_No=338), raw dandelion greens contain 13,610 micrograms of lutein-zeaxanthin per 100 gram (3.5 ounce) serving. If cooked without salt they contain 9,158 micrograms of lutein-zeaxanthin per 100 grams and if cooked with salt they contain 3398 micrograms of lutein-zeaxanthin per 100 grams.

Makes 4 to 6 servings, each of which contains zero carbohydrate units.

Kale with Bacon

Cooking kale with bacon may make it more palatable for those who dislike the taste of kale. However, the "Eye Smoothie," page 87, is probably the best option for making kale taste good.

> 1 bunch of kale weighing ½ to ¾ pound
> 3 strips bacon

Cut the bacon into ½ inch slices and sauté them in a saucepan until crisp. While the bacon is cooking, wash the kale, remove tough or woody stems, and cut into 3-inch pieces. Add to them to the pan with the bacon pieces and rendered fat and wilt them as in the "Greens" recipe on previous page. When they are all wilted, place the lid on the pan and leave it on the burner for 10 or 15 minutes to further soften any remaining stems. Makes 2 to 4 servings containing zero carbohydrate units each.

Braised Brussels Sprouts

These Brussels sprouts are more tasty than boiled sprouts.

> ¾ to 1 pound fresh Brussels sprouts
> 1 to 2 tablespoons olive oil for flavor or other monounsaturated oil such as
>   avocado or canola oil
> Dash of salt or to taste

Wash the Brussels sprouts and cut off the ends if dry or brown. Slice them in half lengthwise. Place 1 tablespoon of the oil in a large frying pan and spread it around to cover the bottom of the pan. Add the Brussels sprouts, cut side down. If they don't all fit in the pan, reserve those that do not fit.

Heat over medium heat until the bottom sides of the sprouts are golden brown. Remove the brownest sprouts to make room for the reserved sprouts. Place the reserved sprouts in the pan cut side down, adding an additional teaspoon to tablespoon of oil if needed. Set the cooked sprouts that were removed from the pan on top of the other sprouts in the pan, uncut side down. Cook until the just-added sprouts are beginning to brown. Place the lid on the pan, reduce the heat, and allow the sprouts to steam for a few minutes. Sprinkle with salt and serve. Makes 3 to 5 servings each of which contains zero carbohydrate units.

Mashed Jersey Sweet Potatoes

Boiled Jersey sweet potatoes have a GI score of 44, well down in the low range. I think they are amazingly delicious mashed with olive oil. However, when I tried to get Mark to eat them instead of mashed white potatoes, he called them "fake mashed potatoes." He liked them better mashed with butter and milk but he hasn't consented to eat them since that turkey dinner. Try them and judge for yourself. I think you'll love them!

> 2 pounds of Jersey (white) sweet potatoes
> 2 to 3 tablespoons butter or monounsaturated oil such as olive oil if
>   avoiding dairy products
> ¼ teaspoon salt or to taste
> ¼ to ½ cup of milk or reserved cooking water if avoiding dairy products

Peel the sweet potatoes and cut them into chunks. Place the chunks in a large saucepan, cover them with water, and bring the pan to a boil. Reduce the heat and simmer for about 20 minutes or until they are tender when pierced with a fork. Drain all of the water, reserving some for mashing if you are not using the milk. Add the butter or oil and about ¼ teaspoon of salt to the pan. Mash the potatoes, adding enough of the reserved cooking water or milk to make them fluffy. Taste them and add a little more salt if needed, until they taste just right to you. Makes 6 servings containing 2 carbohydrate units each. You can make a half batch to yield 6 servings containing 1 carbohydrate unit each, but there is no way I could be satisfied with that small a serving of these yummy potatoes.

Mashed Rutabagas

This recipe was my second attempt to find a substitute for mashed white pota-toes. When I tasted them, I noticed the "rutabaga" flavor and thought these had less chance than the recipe above. However, Mark liked them! We'll see if he likes them with turkey...

> 2 to 2¼ pounds of small to medium size rutabagas
> About 1 teaspoon salt, divided
> 2 to 3 tablespoons butter or monounsaturated oil such as olive oil if
>   avoiding dairy products
> ¼ to ½ cup of milk or alternative milk or water if avoiding dairy products

Choose rutabagas that are not excessively large or old because such roots may never soften well when simmered. Peel the rutabagas and cut them into chunks.

Place the chunks in a large saucepan and cover them with water. Add ½ teaspoon of the salt to the water, place the pan on the stove, and bring the water to a boil. Reduce the heat and simmer for about 30 to 40 minutes or until the rutabagas are tender when pierced with a fork. Drain all of the water and add the butter and about ¼ teaspoon of salt to the pan. Mash the rutabagas, adding enough of the milk to make them fluffy. Taste them and add enough of the remaining salt, or a little more, until they taste just right to you. Makes 4 to 6 servings containing zero carbohydrate units each.

Crock Pot Baked Beans

Our son John likes these beans so much that he makes a 6-quart crock pot batch for himself every two weeks. With the oil and tomato paste, these beans are a good source of lycopene.

> 1 pound small white beans or navy beans
> Water
> ¾ cup apple juice concentrate, thawed, or ¾ cup of water plus ⅛ to
>   ¼ teaspoon Berlin Seeds™ stevia extract or pure monk fruit extract, or
>   to taste
> 1 6-ounce can tomato paste or 1½ teaspoons paprika (optional for color)
> 1 to 3 tablespoons finely chopped onion (optional or to taste)
> 1 teaspoon dry mustard powder (optional)
> 1 teaspoon sweet basil, or more to taste
> 1½ teaspoons salt
> ¼ teaspoon pepper
> 2 tablespoons monounsaturated oil such as olive, avocado or canola
>   (optional, but improves lycopene absorption if tomato paste is used)

The day or evening before you plan to serve this dish, wash the beans by putting them in a strainer and running cold water over them. Remove and discard any shriveled beans. Put the beans in a 3-quart crock pot and fill the pot almost to the top with water. The volume of the water should be two to three times the volume of the beans. The next morning, drain the soaking water from the beans, replace it with fresh water and drain it again two or three times. (This soaking and rinsing process removes indigestible carbohydrates that can cause gas). Pour off all the water after the last rinse. Add 4 cups of water to the crock pot and put the lid on the pot. Cook the beans 4 to 6 hours on high or 7 to 8 hours on low, or until they are quite tender. Check them during cooking and add more water if necessary. It is all right if the level of the water goes a little lower than the level of the beans in the pot.

Stir the tomato paste into the apple juice or into the ¾ cup water until the mixture is smooth. Add this mixture, the oil, the seasonings and stevia or monk fruit extract (if you are using a sweetener extract) to the crock pot. Stir these ingredients into the beans thoroughly. Cover the pot and cook the beans on high another 2 to

3 hours. Check them during cooking and add more water if necessary. If you like a thick sauce, smash a few beans against the side of the pot an hour or so before the end of the cooking time. If the sauce still isn't thick enough, set the lid ajar so some of the liquid can evaporate.

Refrigerate or freeze leftover beans. Makes 7 servings which contain 2 protein units each. If made with the apple juice, each serving also contains 1½ carbohydrate units. The carbohydrates in cooked legumes are mostly indigestible or fiber so this recipe contains zero carbohydrate units if made with stevia or monk fruit extract.

Oven Carrots

This dish will make you a cooked carrot lover, especially if made with home-grown or organic carrots from the farmers' market.

> 2 to 2½ pounds carrots, preferably organic
> About ⅓ cup water, depending on how many carrots are used
> ½ teaspoon salt
> 2 to 3 tablespoons monounsaturated oil such as olive, avocado or canola

Peel or scrub the carrots and cut them lengthwise into halves or quarters if they are large. Lay the carrot sticks parallel to each other in a 2 to 3 quart glass casserole dish with a lid. Add the salt and water and drizzle the oil over the top of the carrots. Cover the dish with its lid and bake the carrot sticks at 350°F for about 1 to 1½ hours, or until they begin to brown and become caramelized. Makes 6 to 8 servings each of which contains zero carbohydrate units.

Oven Purple Cabbage

This method of cooking cabbage brings out its delicious sweet flavor.

> 1 head of purple cabbage weighing about 1½ to 1¾ pounds
> ½ teaspoon salt
> ¼ teaspoon pepper (optional)
> ⅓ cup water
> 3 tablespoons monounsaturated oil such as olive, avocado or canola

Coarsely chop the cabbage and put it into a 3-quart glass casserole dish with a lid. Add the salt, pepper, and water. Drizzle the oil over the top of the cabbage. Cover the dish with its lid and bake at 350°F for 1 to 2 hours, stirring once or twice during the cooking time. Makes 6 to 8 servings each of which contains zero carbohydrate units.

Oven Peas or Beans

Because you start with frozen vegetables, this is quick and easy to prepare.

    1 10-ounce package frozen peas, cut green beans, or lima beans
    ¼ cup water
    ⅛ teaspoon salt
    1 tablespoon monounsaturated oil such as olive, avocado or canola

Combine all of the ingredients in a 1½ quart glass casserole dish. Cover the dish with its lid and bake at 350°F for 1 to 1½ hours for the beans or 20 minutes to 1 hour for the peas. Makes 2 to 3 servings each of which contains zero carbohydrate units.

Easy Baked Sweet Potatoes or Squash

These are the easiest vegetables to have with an oven entrée and are high in carotenoids. Try this recipe with kabocha squash for great flavor.

    White or orange sweet potatoes, each weighing 5 to 5½ ounces, or winter
       squash

Scrub and pierce the potatoes. Cut the squash in half and remove the seeds. Place the potatoes or squash (cut side down) on a baking dish or a baking sheet with an edge. I cover the baking sheet with foil because both types of vegetables can ooze sticky liquid. Bake with the rest of your oven meal for 1 to 2½ hours at 350°F to 450°F or until they are tender when squeezed and your main dish is done. (Use a longer cooking time with lower temperatures). Each potato is two carbohydrate units and the squash contains zero carbohydrate units.

Special Oven Squash

The easiest way to prepare winter squash is in the recipe above. However, if you have time and would like a change from ordinary squash, this is a delicious treat.

    2½ pounds butternut squash
    ¼ teaspoon salt
    2 tablespoons monounsaturated oil such as olive, avocado or canola

Peel the squash. Cut it in half lengthwise and remove the seeds. Slice it into ¼-inch slices. Put the slices into an 11 inch by 7 inch baking dish, sprinkle them with the salt, and drizzle them with the oil. Stir to coat all of the slices. Bake at 350°F for 1½ to 2 hours, turning the slices after the first hour. Makes 4 to 6 servings each of which contains zero carbohydrate units. Leftovers are delicious reheated in the oven or toaster oven.

Crispy Oven Sweet Potatoes

If you want to make a special treat and have time for slicing, these potatoes are delicious as well as being high in vitamin A and carotenoids.

> 2 pounds sweet potatoes (preferably white, but orange are also good)
> 2 tablespoons monounsaturated oil such as olive, avocado or canola
> ½ teaspoon salt
> Pepper or herbs to taste (optional)

Peel or scrub the potatoes and slice them into ¼-inch slices. Put the slices into an 11 inch by 7 inch baking dish, sprinkle them with the salt and optional pepper or herbs, and drizzle them with the oil. Stir to coat all of the slices. Bake at 350°F for 1½ to 2 hours, turning the slices after the first hour. This recipe makes 6 servings which each contain 2 carbohydrate units. Warm leftover potatoes in the oven or a toaster oven to make them crisp and delicious the second time.

Special Ways with Raw Vegetables

Here are two recipes for picnic-ready raw vegetable side dishes. Also serve salads made with dark greens such as spinach and romaine lettuce using the dressing recipes on pages 166 to 168.

Sweet Carrot Slaw

This is a delicious change from coleslaw as well as being high in vitamin A and carotenoids.

> 1 pound carrots
> ¼ cup raisins (optional)
> ¾ cup mayonnaise or "Super Smooth Sauce," both on page 164
> 1 tablespoon apple juice concentrate (optional)
> 1 to 3 tablespoons of water

Wash and peel the carrots. Grate them with a grater or food processor. Combine the carrots and optional raisins in a large bowl. In a small bowl, stir together the dressing, optional apple juice concentrate, and enough water to make a creamy dressing. Add them to the carrot mixture and toss thoroughly until the carrots are completely coated with the dressing. Store in the refrigerator. Makes 6 servings which contain zero carbohydrate units each if the optional ingredients are not used. The raisins add about 5 grams or $1/3$ carbohydrate unit per serving and the apple juice concentrate adds about 2 grams of carbohydrate per serving, for a total of about ½ carbohydrate unit per serving if both raisins and apple juice are used.

Purple Coleslaw

This is the old favorite salad with a new color and additional carotenoids.

¾ to 1 pound of purple cabbage, about half of a large head
1 small carrot, shredded
1 to 2 teaspoons finely minced or grated onion, or to taste (optional)
⅓ cup minced bell pepper, preferably red or orange
¾ to 1 cup mayonnaise, page 164, "Coleslaw Dressing," page 167 or "Super
     Smooth Sauce," page 164
⅛ to ¼ teaspoon salt (optional)
1 to 2 tablespoons lemon juice or ⅛ to ¼ teaspoon unbuffered vitamin C
     crystals or powder
1 to 4 tablespoons water

Wash and core the cabbage. Cut it into wedges and slice each wedge crosswise as thinly as possible or shred it with a food processor. Put the cabbage strips into a large salad bowl. Grate the carrot or shred it with a food processor. Grate the onion and finely chop the pepper. Combine all of the vegetables in a large bowl. In a small bowl, stir together the dressing, salt, lemon juice or vitamin C, and enough water to make a creamy dressing. Add it to the cabbage mixture and toss thoroughly until the cabbage is completely coated with the dressing. Store in the refrigerator. Makes 6 servings which contain zero carbohydrate units each.

Grains and Breads

Bread made from grains is the staff of life. Because grains keep well in storage longer than many other foods, they have made up a large part of our food supply since human beings began growing grain. This chapter includes both true grains that are in the grain family and alternative grains such as quinoa, amaranth, and buckwheat. These can be treated like grain family members in cooking except for the notable difference in quinoa. It is coated with a soap-like substance which protects the seeds from being eaten by insects. Therefore, it must be thoroughly rinsed before cooking or it will have a soapy taste.

The naturopath told Mark to eat whole grains because both grain-family whole grains and alternative grains are highly nutritious. However, she said to completely eliminate white flour and white rice. True grains and grain alternatives are high in vitamin E, B vitamins and a wide variety of minerals including manganese and selenium. (Selenium is especially important for macular degeneration patients). Whole grains are also high in fiber for good digestion and contain some protein, although the level is low compared to legumes or animal foods. Most grains contain incomplete protein, meaning that they do not contain all the amino acids that our bodies are unable to synthesize. Quinoa is an exception to this; it is one of the few plant sources of complete protein, albeit at a low level.

This chapter includes recipes made with a variety of grains to give choices to those on special diets. Macular degeneration patients with wheat allergy or gluten intolerance must strictly avoid the grains which cause problems to prevent causing general inflammation which will aggravate their macular degeneration. For more about this, see page 38.

Grains are the seeds of the plants that produce them. Thus, whole grains contain substances to protect them from being digested by soil bacteria before they can germinate. These enzyme inhibitors and phytic acid also make it less easy for us to digest and absorb nutrients from grains. Phytic acid interferes especially with the absorption of minerals including calcium, zinc, magnesium, iron and copper. However, a diet high in vitamins A, C and D helps prevent adverse effects due to consuming phytic acid. All macular degeneration patients should be eating a diet rich in these vitamins as well as taking them in supplements. This, plus controlling the quantity of carbohydrates eaten, gives some protection from the effects of phytic acid. In addition, healthy intestinal flora can break the phytic acid down.

If poor digestion of whole grains indicates that additional help is needed, whole grains and whole grain flours should be soaked overnight in water with lemon juice or vinegar added to the water before cooking them. Because macular degeneration patients should be "protected" by their supplements and controlled carbohydrate diets, rather than complicate the recipes here, I will direct you to this webpage: http://healingbasics.life/easily-digested-grains-legumes-and-nuts.html or the book *Nourishing Traditions*[1] for more information if needed.

---

1 Fallon, Sally with Mary Enig, PhD. *Nourishing Traditions*., (Brandywine, MD, NewTrends Publishing, 2001).

Fermenting breads with yeast and/or *lactobacilli* (for sourdough) breaks down the phytic acid in the flour, so wheat-containing, non-wheat and gluten-free yeast bread recipes are included in this book. The long fermentation of sourdough bread breaks down phytic acid even more thoroughly and sourdough has a low glycemic index so a few sourdough recipes are included here also.

Macular degeneration patients and families may lack the time, energy or vision to do much baking. Therefore, I suggest (1) eating cooked whole grains (pages 122 to 125) and/or (2) buying 100% whole grain bread. Here, a local bakery chain makes 100% whole wheat bread sweetened with honey, more than just a little honey judging from its flavor. The sweetness of this bread seemed like it would destabilize Mark's blood sugar, and he has never liked the taste of whole wheat. Therefore, the whole wheat sourdough bread recipes on pages 139 to 142 are what he eats. They contain some white bread flour to help them rise but the percentage of white flour is much lower than in commercially made "whole wheat" bread. Because they are sourdough bread, the glycemic index score of these breads is low.

If you can buy 100% whole grain bread that you like and tolerate, simplify your life and purchase rather than bake bread. If you are cooking for an allergy or gluten-free diet and need more recipes, contact me using the information at the bottom of page 4 and I will email you e-books that will provide more recipes.

Take this chapter slow and easy and stop whenever you have had enough. The oven or stovetop cooked whole grains on the next few pages are a good start and can provide all the whole-grain nutrition needed.

Oven Grains

These grains are easy to pop in the oven while the main dish is cooking, yet are so delicious that they make any oven meal special.

Quinoa
> 1 cup quinoa
> 2½ cups water
> ½ teaspoon salt, or to taste
> 1 tablespoon oil
>
> Cooking time: 1 hour; Quinoa should be thoroughly rinsed under running water in a strainer before cooking.

Buckwheat
> 1 cup white or roasted buckwheat groats
> 3½ cups water
> ½ teaspoon salt, or to taste
> 1 tablespoon oil
>
> Cooking time: 1 to 1½ hours

Teff

    1 cup teff
    3 cups water
    ½ teaspoon salt, or to taste
    1 tablespoon oil
Cooking time: 1 to 1½ hours

Brown rice

    1 cup brown rice
    2½ cups water
    1 tablespoon oil
    ½ teaspoon salt
Cooking time: 1 to 1½ hours

Wild rice

    1 cup wild rice
    4 cups water
    1 tablespoon oil
    ½ teaspoon salt
Cooking time: 1½ to 2 hours

Choose one set of ingredients from the list above. If you are preparing quinoa, be sure to rinse it in a strainer under running water until the water is no longer sudsy in order to remove its natural soapy coating. Stir together the grain, water, oil and salt in a 2 to 3-quart glass casserole dish with a lid. Cover the dish and bake at about 350°F until the grain is tender and all the water is absorbed. The baking time is flexible so these grains can usually be baked the same amount of time as the entrée of an oven meal. Approximate baking times for each type of grain are given at the end of each ingredient list above. If you will be baking the grain for much longer than these times or at a higher temperature, you may need to add a little more water. Check the grain near the end of the cooking time the first time you make it and note how much extra water you added, if it was indeed needed. Record how much extra water you used for the next time you cook it at a high temperature or for a longer time. If the grain needs more water at 15 to 30 minutes before mealtime, simply remove it from the oven. The glass casserole dish will keep it warm.

Makes about 2 cups of cooked grain or 4 or more servings. If this size batch is 4 servings of quinoa, each contains 2 carbohydrate units and ¾ protein unit. If it is 4 servings of buckwheat, each contains 2 carbohydrate units and 2/3 protein unit. If it is 4 servings of teff, each contains 2 carbohydrate units and ½ protein unit. If you make brown rice, you will need to divide the recipe into 5 servings; each contains 2 carbohydrate units and ½ protein unit. If it is 4 servings of wild rice, each contains 2 carbohydrate units and 2/3 protein unit.

Stove-Top Grains

Amaranth
  1 cup amaranth
  2½ cups water
  ¼ teaspoon salt
Cooking time: 30 to 35 minutes

Quinoa
  1 cup quinoa (white) or black quinoa[2]
  2 cups water
  ¼ teaspoon salt
Cooking time: 20 minutes for white quinoa, 40-45 minutes for black. Rinse quinoa thoroughly under running water in a strainer before cooking.

Buckwheat
  1 cup buckwheat
  2½ cups water
  ½ teaspoon salt
Cooking time: 20 to 30 minutes

Teff
  1 cup teff
  3 cups water
  ¼ teaspoon salt
Cooking time: 15 to 20 minutes

Brown rice
  1 cup brown rice
  2½ cups water
  ¼ teaspoon salt
Cooking time: 45 to 50 minutes

Wild rice
  1 cup wild rice
  4 cups water
  ¼ teaspoon salt
Cooking time: 60 minutes

---

2 I recently discovered black quinoa in a local health food store. It has a wonderful flavor, slightly sweet with none of the unusual flavor of white quinoa, and is loaded with antioxidants and other nutrients. See Mercola, Joseph, DO. "Are You Curious about Quinoa?" May 11, 2015. https://articles.mercola.com/sites/articles/archive/2015/05/11/quinoa-nutrition.aspx and Curinga, Karen. "Benefits of Black Quinoa," October 3, 2017. https://www.livestrong.com/article/497045-benefits-of-black-quinoa/

Choose one set of ingredients on the previous page. For creamy grains, put all the ingredients in a sauceapan, add the lid and bring it to a boil. Then reduce the heat and simmer for the time specified at the end of the ingredient list.

For fluffy grains, bring the water to a boil in a saucepan. Add the grain and salt. Put the lid on the pan. Allow the water to return to a boil, then lower the heat and simmer for the time specified at the end of the ingredient list. Remove the pan from the heat. Allow the grain to stand for a few minutes, fluff and serve. Makes about 2 cups of cooked grain or 4 or more servings.

If this size batch is 4 servings of amaranth, each contains 2 carbohydrate units and 1 protein unit. If it is 4 servings of quinoa, each contains 2 carbohydrate units and ¾ protein unit. If it is 4 servings of buckwheat, each contains 2 carbohydrate units and ²/₃ protein unit. If it is 4 servings of teff, each contains 2 carbohydrate units and ½ protein unit. If you make brown rice, you will need to divide the recipe into 5 servings; each contains 2 carbohydrate units and ½ protein unit. If it is 4 servings of wild rice, each contains 2 carbohydrate units and ²/₃ protein unit.

Quinoa Pilaf

Quinoa boasts high quality protein and a low GI score of 53. Thus, it is a great substitute for rice. This tasty grain-free side dish is also delicious as poultry stuffing.

> 2 cups sliced celery
> ¼ small onion, chopped (optional)
> 3 to 4 tablespoons oil
> 1 cup quinoa, thoroughly rinsed
> 2 cups water
> ½ to 1 teaspoon salt, or to taste
> ¼ teaspoon pepper
> 1 tablespoon dried parsley
> 1 teaspoon sweet basil
> ¼ teaspoon ground rosemary (optional)

Rinse the quinoa in a strainer under running water until the water is no longer sudsy to remove its natural soapy coating. Sauté the celery and onion in the oil in a saucepan until they just begin to brown. Add the quinoa and water, bring the mixture to a boil, and simmer it for 20 minutes, or until the quinoa is translucent. Stir in the seasonings thoroughly and allow the quinoa to stand for a few minutes so that the flavors can blend. Serve it as a side dish or stuff it into a large chicken and then roast the chicken. If you use this stuffing for a turkey, double the recipe for a 12-pound turkey or triple it for a 24-pound turkey. Makes 4 to 6 servings. If you divide this into 4 servings, each contains 2 carbohydrate units and ¾ protein unit. For 6 servings, each contains 1¹/₃ carbohydrate units and ½ protein unit.

Non-Yeast Baking

A few recipes for baking powder or baking soda leavened baked goods are included here for when whole-grain baked goods are needed quickly. This section will also acquaint you with allergy and gluten-free baking in case you have sensitivities that you must take seriously for the sake of your eyes.

Here are non-yeast baking basics: The first thing to do is to purchase the correct ingredients. High quality flour that is consistent from bag to bag will save you money, time and trouble in the long run. If you are leavening your baked goods with baking soda plus vitamin C[3], be sure that the vitamin C powder or crystals that you use are unbuffered. If you are allergic to corn, also make sure they are corn-free. Your baking powder should be gluten-free and/or corn-free if you are gluten-sensitive or corn-allergic. (See "Sources," pages 212 and 217, for information about corn-free baking powder and tapioca-source unbuffered vitamin C). Then measure the ingredients accurately. Fluff the flour with a spoon, scoop it into the measuring cup(s) and level it off with a knife.

Proper mixing is the special diet baking technique that is most crucial for success. Because the types of flour used in special diet baking produce a more fragile structure which must trap the gas produced by the leavening, how and how much you mix non-yeast baked goods and how quickly you get them into the oven is very important. Before you begin baking, preheat the oven. Prepare the baking pan(s) ahead of time; oil and flour them with the same kind of flour used in the recipe. Mix the dry ingredients together in a large bowl and the liquid ingredients together in the measuring cup or another bowl. Working quickly, stir the liquid ingredients into the dry ingredients until they are **just mixed.** It is critical that you do not mix for too long or the leavening will produce most of its gas in the mixing bowl rather than in the baking pan in the oven. Only stir until the dry ingredients are barely moistened; a few floury spots in your batter or dough are all right. As soon as you have mixed just enough, quickly put the batter or dough into the prepared baking pan(s) and pop them into the preheated oven.

To test quick breads, muffins, etc. for doneness, insert a toothpick into the center of the pan. If it comes out dry, it is time to remove the pan from the oven. Most baked goods should be removed from the pan immediately after you take them from the oven, but some fragile items benefit from cooling in the pan for about 10 minutes. Allow your baked goods to cool completely on a cooling rack before you slice them or store them. While plastic bags and containers are great for storing most baked goods, crackers will stay crisp if you store them in a metal tin.

For more information about vitamin C, allergy and gluten-free baking or more recipes, contact me using the contact information at the bottom of page 4.

---

3 Unbuffered vitamin C is used as an acid ingredient for allergy cooking because (1) it is highly purified and (2) baking powder usually contains cornstarch. To prevent sensitization to the potato starch in corn-free baking powder, it should not be used on a daily basis. Cream of tartar is giving as an alternate in the baking recipes here, but if you are allergic, for your own sake, purchase a bottle of the vitamin C on page 217 which will last a long time.

Banana Bread

This sweet, moist bread can be made in either gluten-free non-grain (amaranth) or grain-containing versions.

Choose one type of flour or flour-starch combination:
    3¼ cups Purity Foods™, Arrowhead Mills™ or Bob's Red Mill™ whole
        spelt flour
    2½ cups barley flour
    2 cups amaranth flour plus ½ cup arrowroot or tapioca flour
½ teaspoon salt
2 teaspoons baking soda
½ teaspoon unbuffered vitamin C powder OR 1 tablespoon cream of tartar
½ teaspoon ground cloves or 1 teaspoon cinnamon (optional)
½ cup chopped nuts (optional)
1¾ cups pureed or thoroughly mashed **ripe** bananas
¼ cup oil

Preheat your oven to 350°F. Oil and flour (using the same flour you are baking with) a 9 by 5 inch loaf pan. Stir together the flour(s), baking soda, vitamin C powder or cream of tartar, spice, salt, and nuts in a large bowl. Thoroughly combine the pureed bananas and oil and stir them into the dry ingredients until they are mixed in, but *be careful not to over-mix.* A few floury spots in the dough are all right. (The batter will be stiff). Scrape the batter into the prepared loaf pan and bake for 55 to 60 minutes, or until the bread is lightly browned and a toothpick inserted in the center comes out dry. Remove it from the oven and allow it to cool in the pan for 10 minutes. Remove it from the pan to cool completely. Makes one loaf. For spelt banana bread, slice it into 11 slices containing 2 carbohydrate units each or 15 slices containing 1½ carbohydrate units each. For barley banana bread, slice it into 12 slices containing 1 carbohydrate unit each or 8 slices containing 1½ carbohydrate units each. For amaranth banana bread, slice it into 9 slices containing 1½ carbohydrate units each.

Whole Wheat Blueberry Muffins

These muffins are delicious as well as being quick and easy to make.

2¼ cups stone ground whole wheat flour such as Bob's Red Mill™ brand
3 teaspoons baking powder
¼ teaspoon salt
⅓ cup water
⅔ cup apple juice concentrate, thawed
⅓ cup oil
1 cup fresh or frozen blueberries

Preheat your oven to 350°F. Oil and flour 12 muffin cups or line them with paper liners. If you are using frozen blueberries, put them in a strainer, run cold water over them and allow them to drain while making the batter.

In a large bowl mix together the flour, baking powder, and salt. Combine the water, apple juice and oil in a separate bowl or measuring cup. Before the liquid ingredients can separate into layers, quickly stir them into the dry ingredients until just mixed. Gently but quickly fold in the blueberries. Fill the muffin cups 7⁄8 full. Bake for 20 to 24 minutes or until a toothpick inserted in the center of the largest muffin comes out dry. Remove the muffins from the pan and allow them to cool on a wire rack. Makes 12 muffins each of which contains 1½ carbohydrate units.

Quinoa Sesame Seed Crackers

These make great snacks, especially with cheese. You won't miss the crunch of processed food snacks when you have these.

    3 cups quinoa flour
    1 cup tapioca starch or arrowroot
    ⅜ cup sesame seeds
    2 teaspoons baking soda
    ½ teaspoon unbuffered vitamin C powder OR 1 tablespoon cream of tartar
    1 teaspoon salt
    1¼ cups water
    ½ cup oil
    Additional salt for the tops of the crackers

Preheat your oven to 350°F. Mix together the quinoa flour, starch, sesame seeds, baking soda, vitamin C powder or cream of tartar, and 1 teaspoon of salt in a large bowl. Whisk together the water and oil and quickly, before they separate, stir them into the dry ingredients until the dough roughly comes together. Divide the dough into halves or thirds. (If your baking sheets are 14 by 18 inches or larger, you can bake these crackers using two sheets; otherwise divide the dough into thirds). Press each portion of the dough together on the ungreased baking sheet and sprinkle the top with quinoa flour. Roll the dough to a little over ⅛ inch thickness using an oiled rolling pin. If the dough sticks to the rolling pin, lightly flour the top of the dough again. After rolling, cut the dough into 1½ inch squares and sprinkle the tops of the crackers with salt. Bake for 15 to 25 minutes, or until the crackers are crisp and lightly browned. The crackers around the edges of the baking sheet will probably brown before those in the center. Remove them from the baking sheet and allow the crackers in the center to bake 5 minutes longer. Then remove any others that are brown and allow the remaining crackers to bake 5 minutes longer again. Makes about 9 dozen crackers or 21 5-cracker servings. A serving of 5 crackers contains 1 carbohydrate unit.

Pineapple Muffins

Delicately flavorful, these muffins are a hit with young and old alike especially when made with sorghum for those who avoid gluten.

> Chose one type of flour:
>   2½ cups Purity Foods™, Arrowhead Mills™ or Bob's Red Mill™ whole
>   spelt flour
>   2 cups sorghum flour
>   2 cups teff flour
> 1 teaspoon baking soda
> ¼ teaspoon unbuffered vitamin C powder OR 1 teaspoon cream of tartar
> ½ cup pineapple canned in juice or fresh pineapple with juice to cover
> ½ cup pineapple juice concentrate, thawed
> ⅓ cup oil

Preheat your oven to 400°F. Line 12 wells of a muffin pan with paper liners (which are especially helpful for the fragile sorghum and teff muffins) or oil and flour them with the flour you are using in the recipe.

Mix together the flour, baking soda, and vitamin C powder or cream of tartar in a large bowl. Puree the pineapple and its juice with a hand blender, standard blender, or food processor. Thoroughly combine the pureed pineapple, pineapple juice concentrate and oil and stir them into the dry ingredients until they are just mixed in. Put the batter into the prepared muffin tin, filling the cups about 2/3 full. Bake for 15 to 20 minutes or until the muffins begin to brown and a toothpick inserted in the center of the largest muffin comes out dry. Makes about 12 muffins each of which contains 1½ carbohydrate units.

About Yeast Bread

The best and easiest way to make yeast bread is to use an appliance such as a heavy duty mixer (like a KitchenAid™) or a bread machine. Therefore, separate detailed instructions for making bread totally by hand are not included in this chapter. If you have made yeast bread by hand, the whole wheat or spelt bread recipes on page 131 or 134 contain brief hand instructions that will allow you to make bread using the traditional method you are familiar with. The gluten-free bread recipes in this chapter must be made using a mixer or bread machine to develop the structure of the guar or xanthan gum.

For information on how to make yeast bread by hand, mixer, or bread machine, see the third edition of *Easy Breadmaking for Special Diets* as described on page 229 or here: http://www.healingbasics.life/eb.html . For up-to-date (at the time of this writing) information about choosing the bread machine best for you, turn to pages 206 to 209 of this book. If you are the cook for a macular degeneration patient and would like more information or more bread recipes, contact me using the information at the bottom of page 4 for a no-charge e-book of *Easy Breadmaking*.

If you already have a bread machine, you may be able to use it for special breads. Almost any machine can be used on the dough cycle to perform the initial mixing and kneading and the first rise for your bread. Then restart the cycle and allow the dough to knead for another 3 to 5 minutes. Remove the dough from the machine and put it in a prepared loaf pan. Proceed with the second rise and baking as directed in the recipe. If your machine has a bake-only cycle, you may also be able to bake it in the machine. See the fouth paragraph on page 206 for how to do this.

If your kitchen does not have a draft-free warm spot for bread dough to rise, you can make one in the oven. If you have a gas oven, the pilot light will probably keep the oven at just about the right temperature, 85 to 90°F. If you have an electric oven, turn it to 350°F and let it preheat. Turn it off, open the oven door, and let it cool for about 10 minutes until the temperature inside the oven is about 85 to 90°F. Then close the oven door. Use a thermometer to verify that the oven is at the correct temperature the first few times you do this. After you gain experience, you will be able to tell the right temperature by the feel of the inside of the oven when you place dough in it to rise.

When you use your oven as a cozy place to keep a sourdough sponge warm for 18 to 20 hours or to keep sourdough dough warm and rising for several hours, you will need to take action occasionaly to maintain the temperature at about 85 to 90°F. Monitor the temperature inside the oven every hour or two with at thermometer. If the oven is too cool, turn it on for 20 to 30 seconds. Then put the digital thermometer in the oven for five minutes or so while the temperature equilibrates. Read the thermometer to see if the oven is back in the 85 to 90°F range, and repeat a brief heating if not. As you work with your oven, you will learn how long to heat it to produce the amount of rise in temperature needed.

Gluten-free bread recipes in most cookbooks usually are made with a combination of several flours plus multiple stabilizers. This seems to produce bread with a more conventional texture in most cases. However, I do not combine several grains in each recipe for a few good reasons: (1) to save money on many stabilizing ingredients, (2) to save time on measuring, which makes bread making go more quickly and easily and (3) most importantly, to prevent the development of allergies to foods that are eaten every day such as the stabilizers and the foods they come from. Therefore, most of the recipes in this book are made with a single grain/grain alternative or a single grain/grain alternative plus a starch[4] which acts as a binder. My most fool-proof and favorite single-grain-plus-starch gluten-free yeast bread recipe is the buckwheat bread recipe on page 135. For more bread recipes, both made by hand and by bread machine, see *Easy Breadmaking* mentioned above.

---

4 If baking for food allergies, always use the same starch with each type of non-grain flour to preserve tolerance. For example, use tapioca with quinoa, arrowroot with amaranth, bean starch with buckwheat, etc. This allows for a set rotation of both the grain and the starch.

Oven-Baked 100% Stone Ground Whole Wheat Bread

Bread made from only stone ground whole wheat flour has a lower glycemic impact than mostly-white-flour "whole wheat" bread from the grocery store.

1⅛ cups warm water
1 tablespoon single source honey or agave (to feed the yeast)
1½ tablespoons oil or 1 tablespoon oil plus ½ tablespoon liquid lecithin
1 teaspoon salt
3 cups stone ground whole wheat flour plus additional flour if needed
3 tablespoons vital gluten such as Bob's Red Mill™ brand
2¼ teaspoons (1 packet) active dry yeast

This bread may be prepared by hand, with a mixer, or using the dough cycle of a bread machine. (See the next recipe for the bread machine option). Mix the vital gluten into the flour thoroughly before proceeding with any of these methods.

To use a mixer, put about 2 cups of the flour-gluten mixture, the yeast and the salt in the mixer bowl. Mix on low speed for about 30 seconds. Warm the liquid ingredients to 115 to 120°F. With the mixer running on low speed, add the liquids to the dry ingredients in a slow stream. Continue mixing until the dry and liquid ingredients are thoroughly mixed. If your mixer is not a heavy-duty mixer, at this point beat the dough for 5 to 10 minutes. You will be able to tell that the gluten is developing because the dough will begin to climb up the beaters. Then knead the rest of the flour in by hand, kneading for about 10 minutes, or until the dough is very smooth and elastic.

If your mixer is a heavy-duty mixer, very little hand kneading is required. After the liquids are thoroughly mixed in, with the mixer still running, begin adding the rest of the flour around the edges of the bowl ¼ cup at a time, mixing well after each addition before adding more flour, until the dough forms a ball and cleans the sides of the bowl. Knead the dough on the speed directed in your mixer manual for 5 to 10 minutes or until the dough is very elastic and smooth. If it softens during kneading, add more flour. Turn the dough out onto a floured board and knead it briefly to check the consistency of the dough, kneading in a little more flour if necessary.

Put the dough into an oiled bowl and turn it over so that the top of the ball is also oiled. Cover it with plastic wrap and then with a towel and let it rise in a warm place (85°F to 90°F) until it has doubled in volume, about 45 minutes to 1 hour. (See the second paragraph on page 130 for how to make a cozy spot for your handmade or mixer dough to rise).

While the dough is rising, oil a loaf pan that is, if possible, 4 to 4 ½ inches wide and 8 to 10 inches long.[5] When the dough has doubled in size, remove the dough

---

[5] Norpro makes a 10 inch long, 4.5 inch wide loaf pan that is ideal for achieving the right serving sizes for a glycemic control diet. The narrow slices of bread it produces make it possible to slice the loaf more thickly without exceeding about 1½ ounces per slice of bread, or one 15 gram carbohydrate unit serving per slice. See "Sources," page 211, for where to get this pan.

from its rising place, knead it down briefly, and form it into a loaf. Place it in the prepared pan. Let it rise in a cozy place for 30 to 45 minutes or until it is a little more than doubled; then remove it from the oven if that is its cozy rising place. Preheat oven to 350°F. Put the loaf into the oven and bake it for about 45 minutes, covering it with foil after the first 30 minutes to prevent excessive browning. Remove the loaf from the pan. It should sound hollow when tapped on the bottom with your knuckles. Allow the loaf to completely cool on a wire rack before slicing it. Makes one 1½ pound loaf, or about 16 1½-ounce slices each of which contains 1 carbohydrate unit. These slices are perfect for staying under two carbohydrate units in a meal containing a sandwich. If you wish to have thicker slices and eat only one per meal, slice the loaf into 11 slices each of which contains 1½ carbohydrate units.

*Note on the second rise: When I let this loaf rise in the bread pan, my goal is to produce the tallest loaf I can without letting it rise for so long that it will fall. For my pans, I allow it to rise until it is about ¾ inch above the top of the pan; the time that this takes varies. This is probably well over twice the volume of the original tightly wrapped, totally deflated roll of dough I put in the pan. For taller slices, you may want to experiment to see how much you can let your loaf rise without falling.

Bread Machine 100% Stone Ground Whole Wheat Bread

This whole wheat bread is delicious and near-effortless to make with almost any bread machine. However, due to the high level of fiber it contains, the middle of the loaf may fall a little.

> 1⅛ cups warm water
> 1 tablespoon single source honey or agave (to feed the yeast)
> 1½ tablespoons oil or 1 tablespoon oil plus ½ tablespoon liquid lecithin
> 1 teaspoon salt
> 3 cups stone ground whole wheat flour plus additional flour if needed
> 3 tablespoons vital gluten such as Bob's Red Mill™ brand
> 2¼ teaspoons (1 packet) active dry yeast

Mix the vital gluten into the flour thoroughly. Then add the ingredients to the pan in the order listed in your bread machine's instruction manual. (For most machines this will be the order given above). Chose a loaf size of 1½ pounds and start the basic cycle or a quick wheat cycle* if your machine has one. (I use the quick wheat cycle on my Zojirushi machine). Start the machine. After 5 to 10 minutes of vigorous kneading, touch the dough. If it is sticky, add 1 tablespoon of flour. After a few more minutes of mixing, re-check the consistency of the dough and if it is still sticky, add another 1 tablespoon of flour. If it is supple but not sticky, you have added enough flour. If you live in a humid climate and your flour absorbs moisture from the air, you may need to add a total of 1 to 3 additional tablespoons of flour to the dough to achieve the right consistency – supple but not sticky. In a dry climate during a dry season, you may need to add 1 to 3 teaspoons of water (one at a time) to make the dough supple.

Allow the cycle to run to completion. Remove the bread from the machine and allow the loaf to completely cool on a wire rack before slicing it. Store it in a plastic bag. Slice it thinly into 16 slices. Makes one 1½ pound loaf, or about 16 1½-ounce slices each of which contains 1 carbohydrate unit. Thin slices are perfect for sandwiches because the sandwich will contain 2 carbohydrate units, the limit per meal on a glycemic control diet. If you wish to have thicker slices and eat only one per meal, slice the loaf into 11 slices each of which contains 1½ carbohydrate units.

*Note on the cycle to use with this bread recipe: Because this bread has a high fiber content, it may fall a little. If this happens with the first cycle you try on your bread machine, switch to another cycle that has a shorter last rise time. If you have a programmable machine, try using a cycle similar to this: Knead – 20 minutes; Rise 1 and 2 – Off; Rise 3 (or the last rise) – 30 minutes; Bake – 60 minutes. Be prepared to tweak this cycle to fit your machine and baking conditions such as climate and elevation. If the loaf over-rises and falls, decrease the time of the last rise. If it comes out dense, increase the last rise progressively – such as 5 minutes each time you make the bread – until it falls in the middle. Then decrease it to the previous setting. Another possible help for the problem of over-rising and then falling is to add an extra ½ tablespoon of gluten. Experiment with the various cycles on your machine, especially any that have a short (30 minute or less) last rise. The whole wheat cycle on most machines includes a lot of rising time, which is not what you want when your bread over-rises and falls. The basic cycle for white bread is more likely to work well.

Whole Spelt Yeast Bread

Spelt really shines in yeast bread. This loaf is so normal in taste and texture that guests may be surprised that they are not eating wheat.

> 1 cup water
> ¼ cup apple juice concentrate, thawed
> 1½ tablespoons oil
> 1 teaspoon salt
> About 3½ to 3⅞ cups Purity Foods™ or Arrowhead Mills™ whole spelt
>   flour** plus additional flour as needed during kneading
> 2¼ teaspoons (1 packet) active dry yeast

To make this recipe with a bread machine, add the ingredients to the pan in the order listed by your bread machine's instruction manual using the smaller amount of the flour. (For most machines this will be the order given above). Chose the basic cycle and a loaf size of 1½ pounds. Start the machine. After a few minutes of mixing, look in the machine. If the dough is very soft, begin adding more flour about 2 tablespoons at a time until it reaches a consistency that is not sticky. After about 10 minutes of kneading, re-check the consistency of the dough and add flour if needed. It should not be soft or sticky and should be starting to become elastic. Re-check the dough near the end of the kneading time, such as when the "add raisins" timer sounds, because spelt dough can soften during kneading and may need more flour. Then allow the rest of the cycle to run to completion.

To make this recipe with a mixer, put about $2/3$ to $3/4$ of the flour ($2\frac{1}{4}$ to $2\frac{3}{4}$ cups of the flour for this recipe), the yeast, and the salt in the mixer bowl. Mix on low speed for about 30 seconds. Warm the liquid ingredients to 115 to 120°F. With the mixer running on low speed, add the liquids to the dry ingredients in a slow stream. Continue mixing until the dry and liquid ingredients are thoroughly mixed. If your mixer is not a heavy-duty mixer, beat the dough for 5 to 10 minutes. You will be able to tell that the gluten is developing because the dough will begin to climb up the beaters. Then knead the rest of the flour in by hand, kneading for about 10 minutes, or until the dough is very smooth and elastic. If your mixer is a heavy-duty mixer, after the liquids are thoroughly mixed in, with the mixer still running, begin adding the rest of the flour around the edges of the bowl about ¼ cup at a time, mixing well after each addition before adding more flour, until the dough forms a ball and cleans the sides of the bowl. Knead the dough on the speed directed in your mixer manual for 5 to 10 minutes or until the dough is very elastic and smooth. If it softens during kneading, add more flour. Turn the dough out onto a floured board and knead it briefly to check the consistency of the dough, kneading in a little more flour if necessary. Put the dough into an oiled bowl and turn it over so that the top of the ball is also oiled. Cover it with plastic wrap and then with a towel and let it rise in a warm (85°F) place[6] until it has doubled in volume, about 45 minutes to 1 hour.

While the dough is rising, prepare your baking pan. Spelt is different from other types of bread in that it can be very difficult to get the bread out of the pan. Rub the inside of an 8-inch by 4-inch or 9-inch by 5-inch loaf pan with oil. Cut a piece of parchment paper the length of the pan and put it in the pan so the bottom and sides are covered with the paper. Oil the paper also. When the dough has doubled in volume, punch it down and shape it into a loaf. Put the loaf into the prepared loaf pan. Allow it to rise until double again. Bake bread at 375°F for 45 minutes to an hour or until it is nicely browned. Check it midway through the baking time, and if it is already getting brown, cover it with a piece of foil to prevent over-browning. At the end of the baking time, remove the loaf from the pan. You may need to run a knife along the ends of the pan which are not lined with paper to loosen the loaf. When it is done, the loaf should sound hollow when tapped on the bottom. Cool it completely on a cooling rack before slicing it with a serrated knife.

Makes one loaf or 12 slices of bread each of which contains 1½ units of carbohydrate. If you want 1-carbohydrate unit slices for sandwiches, use a long, narrow Norpro™ bread pan (see "Source," page 211) and divide the loaf into 17 slices.

*Note on bread machines: Spelt bread will not come out well if made with a bread machine which over-kneads the bread. The gentle kneading of Zojirushi machines is ideal for spelt.

---

6 See page 130, the second paragraph, for how to create a good rising place in your oven.

Note on spelt flour: No matter what brand of flour is used, spelt flour can vary from day to day and from batch to batch on how much flour it takes to make dough of the right consistency. Most bread machine recipes do not require much adjustment, and gluten-free recipes should be made "as written" the first time, because the best consistency of gluten-free dough varies with the grain used. However, in this recipe, you should check the dough several times throughout the kneading time and may need to add more flour because spelt bread softens as it kneads.

Buckwheat "Rye" Yeast Bread

The caraway seeds and rye flavor powder give this bread a delicious rye-like taste. If you have rye fans at your house, watch out! I have to freeze part of the loaf to keep the whole loaf from vanishing in a day or two. If you dislike rye or are allergic to the corn the flavor contains, this bread is delicious without the rye flavor.

½ cup water
¼ cup apple juice concentrate
About 4 large or 3 extra large eggs*, or enough to measure ¾ cup in volume, at room temperature
3 tablespoons oil
1¼ teaspoon salt
1 tablespoon caraway seed (optional)
¾ teaspoon Authentic Foods™ rye flavor powder, to taste (optional – This is gluten free but contains corn. See "Sources," page 214).
1 tablespoon guar or xanthan gum** (See "Sources," page 215).
2 cups buckwheat flour preferably Bob's Red Mill™ flour
1⅜ cup tapioca starch or arrowroot
2¼ teaspoons (1 package) active dry yeast

If you wish to use a non-programmable bread machine to do part of the work by mixing the dough for this bread, see the first full paragraph on page 130. After the first rise, re-start the cycle and mix the dough for 5 minutes. (If your machine mixes slowly at first, begin timing this after it switches to kneading more quickly). Then proceed with this recipe as in the third paragraph of these directions below.

To make this bread using a mixer, heat the water and apple juice concentrate to about 115°F. Beat the eggs slightly and measure ¾ cup, adding a little water to bring the volume up to ¾ cup if needed. Stir together the dry ingredients in a large electric mixer bowl. With the mixer running at low speed, gradually add the juice mixture, eggs and oil. Beat the dough for three minutes at medium speed. Scrape the dough from the beaters and the sides of the bowl into the bottom of the bowl. It will be sticky. Oil the top of the dough and the sides of the bowl. Then cover the bowl with plastic wrap and a towel. Put the bowl in a warm (85°F to 90°F) place*** and let the dough rise for 1 to 1½ hours or until double. Beat the dough again for three minutes at medium speed.

Oil and flour an 8 by 4 inch loaf pan. Put the dough in the pan and allow it to rise in a warm place*** for about 20 to 35 minutes, or until it barely doubles. Preheat the oven to 375°F. Bake the loaf for about 50 to 65 minutes, loosely covering it with foil after the first 30 to 45 minutes if it is getting excessively brown. Remove the loaf from the pan and cool it completely on a cooling rack. Makes one 3-inch tall loaf or 15 slices of bread each of which contains 1½ units of carbohydrate. If you want thicker 2-carbohydrate unit slices, divide the loaf into 11 slices.

This bread can also be made from start to finish in a programmable machine using the following cycle: Knead – 20 minutes; Rise 1 and 2 – Off; Rise 3 (or the last rise) – 25 minutes; Bake – 57 minutes. Be prepared to tweak this cycle to fit your machine and baking conditions. If the loaf over-rises and falls, decrease the time of the last rise. (See pages 206 to 209 for more information about bread machines).

*Note on eggs: If you are allergic to eggs, use ¾ cup warm water in their place. If you do not take the eggs out of the refrigerator early enough for them to come to room temperature before you are ready to bake, put them in a bowl of warm water for 5 or 10 minutes before cracking them to use in this recipe.

**Note on guar or xanthan gum: When mixing the dough for this bread in a bread machine, mix the guar or xanthan gum into the flour before adding them to the machine. If you just add the ingredients to the machine in the order listed, during the warm-up time before mixing begins in some machines, the water and gum can form lumps that routine mixing may not completely eliminate.

***Note on a warm rising place: To use your oven as a cozy place for bread to rise, see the second full paragraph on page 130.

About Sourdough Bread

Sourdough is yeast bread that is leavened by a sourdough starter or culture. Traditional cultures contain wild yeast, which produces gas and causes the bread to rise, and bacteria of the genus *Lactobacillus,* which give the bread a sour flavor and lower its glycemic index score. There are many different traditional sourdough cultures, each with a special flavor of its own and unique rising characteristics. Traditional cultures are sold "live" or freeze-dried in a small amount of wheat dough.

The use of traditional sourdough cultures may be too much work for many with health problems. However, some people who are allergic to commercial baker's yeast and the bread made with it seem to tolerate sourdough bread. Sourdough bread is not yeast-free; perhaps these people are not allergic to the wild yeast but are allergic to baker's yeast much as one may be allergic to lettuce but not to endive. If you are allergic to yeast, you may want to ask your doctor about trying sourdough bread. See *Easy Breadmaking for Special Diets,* 3ʳᵈ Edition, if you wish to bake with traditional sourdough cultures. For more about *Easy Breadmaking* see the last pages of this book.

For macular degeneration patients, the most important reason to make sourdough bread is because it has a lower glycemic index (GI) score than bread made with the same grain but leavened with yeast only. The acid produced in bread by *lactobacilli* decreases the bread's impact on blood sugar and insulin levels and thus makes sourdough bread good for those who need to keep blood sugar levels stable.

A final reason to make your own sourdough bread is for the flavor of the bread itself. You may consider the time spent making sourdough worthwhile (especially if the process is made easier by using a freeze-dried starter) when you taste how delicious very fresh sourdough bread can be.

Making Bread with a Freeze-Dried Sourdough Starter

In the last several years, new products have become available which enable us to make sourdough bread easily and without keeping and maintaining a traditional sourdough starter. The most important of these products is the Florapan™ freeze-dried sourdough starter which is sold by King Arthur Flour ™ in their catalogue and on their website. In addition, bread machines have become more sophisticated than in the past, and with the programmable Zojirushi™ machines we can make this next generation sourdough bread totally in the machine. Both the bread machine (Zojirushi™ Home Bakery Supreme, model BBCEC20) and sourdough starter are available from King Arthur Flour™. See "Sources," pages 211 and 216.

The Florapan™ freeze-dried starter makes assertively sour bread which Mark says is "just like San Francisco sourdough." This starter is gluten-free and wheat-free (but may contain traces of beef) so bread made using it is truly wheat and gluten-free sourdough bread. Unlike breads made with traditional starters, bread made with this starter rises predictably from batch to batch, which allows us to use a programmable bread machine for the whole sourdough process.

The **procedure for making sourdough bread using a programmable bread machine and a freeze-dried starter** is as follows: In the late morning or early afternoon of the day before you plan to serve sourdough bread for dinner, combine the sponge ingredients above the line in the recipe, which are usually flour, purified water (chlorine-free, either bottled water or water purified by a home water filter) and the Florapan™ starter. Thoroughly mix the sponge ingredients using your choice of (1) a wooden or plastic spoon in a glass or ceramic mixing bowl, (2) a heavy-duty mixer or (3) a bread machine's dough cycle or this programmable cycle:

Knead 1 – 15 minutes; Rise 1 – 23½ hours, Rise 2 and 3 – off, Bake – off.

Allow this sponge to rise in a cozy spot (85° F) for 18 to 20 hours in the machine on the programmed cycle above or as described on page 130.

The next morning, add the dough ingredients listed below the line in the recipe you are using. Start the second day cycle given in the recipe. For wheat and spelt bread, adjust the consistency of the dough during the "knead" time. At the end of the knead time, use a spatula to level the dough in the pan. (This makes the slices of bread more uniform in size). Let the cycle run to completion. If the bread is less than ideal, the next time you make it, adjust rising and baking times slightly (start with 5 minute changes) for your baking conditions. If the dough over-rises and falls, decrease the rising time; if it is too dense, increase the rising time; if it over browns, decrease the baking time. Record the times you used for future use.

G**luten-free sourdough bread** may be made using any mixer as in the recipe on page 135 and baked in the oven. Gluten-free bread and most types of wheat-free allergy bread (except spelt and rye) must be mixed with a mixer or bread machine to develop the "structure" of the guar or xanthan gum which is used to simulate the function of gluten.

To make gluten-free sourdough bread using a programmable machine, the procedure for the first day is the same as on the previous page. On the second day, stir the guar or xanthan gum into the flour. Add the flour-gum mixture and the rest of dough ingredients to the sponge. Start the second day cycle given in the recipe. You will see fibrous strands of guar or xanthan gum developing in the thick batter as the kneading progresses. During the kneading time, use a narrow spatula to assist the kneading process. Use the spatula to spread the dough evenly in the pan at the end of the kneading time. Let the cycle run to completion.

The major difference between wheat bread and gluten-free or most allergy breads is that **the best consistency for each type of special bread is different depending on the grain used. Therefore, you should not try to adjust the consistency of the dough during kneading the first time you make these types of bread. Follow the recipe and measure accurately**. However, adjustment may be needed for factors such as humidity, altitude, etc. Changes can be made with the next time you bake this bread, and the first loaf will most likely still be delicious.

If the top of the first loaf has fallen slightly, decrease the last rising time by 10% the next time you make the bread. Make other changes as described in the paragraph at the bottom of the previous page.

For more help, email me using the contact information at the bottom of page 4, and I will send you a no-charge e-book of *Easy Breadmaking for Special Diets*[7] which contains full sourdough information and recipes. The book also contains many other allergy, gluten-free and *normal* bread recipes.

This chapter contains very few sourdough recipes, only one of which is gluten-free, because sourdough is labor intensive and I felt that most macular degeneration patients and their families would not use more that the most predictable basic sourdough recipes. If you are "the cook" and need help or more bread recipes of any kind, please contact me for advice and a no-charge e-book using the contact information at the bottom of page 4.

Although sourdough bread provides great nutrition, yeast bread is much easier to make and has also had the enzyme inhibitors and phytic acid of the grain neutralized by the yeast fermentation. These substances that bind minerals and other nutrients can also be neutalized by soaking whole grains overnight before cooking them.[7] The cooked whole grains and grain alternatives in the recipes on pages 122 to 125 are "back to basics" in work and nutrition as well as delicious. If making bread is too much, enjoy cooked whole grains instead.

---

7 Soaked grain recipes with helpful acidic ingredients are found in *Healing Basics*. Read about this book and *Easy Breadmaking for Special Diets* on pages 227 and 229 or here: http://healingbasics.life/hb.html and http://healingbasics.life/eb.html

Whole Wheat Sourdough Bread

This is the bread Mark eats most of the time. It is on the dense side, but he likes the flavor when made with white whole wheat flour. If you're a whole wheat fan, make it with traditional whole wheat flour, which produces a less dense, taller loaf.

Ingredients for 1 loaf:	Amount
King Arthur™ whole wheat flour OR white whole wheat flour	3 cups
Vital gluten – stir into whole wheat flour	⅜ cup
Florapan™ starter	¼ teaspoon
Non-chlorinated water, 85°F	2 cups
---	---
King Arthur ™ bread flour	2⅛ cups with whole wheat flour OR 2 cups with white whole wheat flour (+ up to 6 additional tablespoons to adjust dough consistency)
King Arthur™ whole grain bread improver	2 tablespoons
Berlin Seeds™ stevia extract[8]	Optional - range ⅛ to ½ teaspoon
Sea salt[9] (preferably unrefined)	1½ teaspoons
SAF™ instant yeast	2 teaspoons
Apple juice concentrate	2 tablespoons
Non-chlorinated water, 85°F	0 to 2 tablespoons for final adjustment

Cycle for day 1 and for mixing on day 2: Dough cycle or this programmable cycle: Knead 1 – 15 minutes; Rise 1 – 23 to 24 hours, Rise 2 and 3 – off, Bake – off.

Cycle for day 2 for bread machine loaf: Knead 1 – 20 minutes, Rise 1 – 3½ hours (range of about 3¼ to 3¾ hours), Rise 2 and 3 – off, Bake – 1 hour, 10 minutes.

In the late morning or early afternoon of the day before you want sourdough bread, begin by stirring the ⅜ cup of vital gluten into the whole wheat flour. Then mix together all of the sponge ingredients above the line in the ingredient list using a wooden or plastic spoon in a glass or ceramic mixing bowl. or using a mixer, a bread machine's dough cycle, or the day 1 programmed cycle above. If using a bread

---

8 I have used only stevia in this recipe, but pure monkfruit extract could probably be substituted in the same amout or to taste. Now that he is used to whole wheat, Mark no longer thinks a sweetener is needed in this bread because of the excellent sourdough flavor.

9 This bread rises better if the fine grind of Celtic™ sea salt is used in this recipe. This is an unrefined salt and contains all the minerals in the ocean. Yeast obviously benefits from trace minerals as much as macular degeneration patients do! Do not be put off by the gray color of the salt.

machine, use a rubber spatula to make sure all the flour is mixed with the water at the very beginning of the kneading time. Allow this sponge to rise in a cozy (85° F) place in your kitchen covered with a towel, in your oven (see the second and third paragraphs on page 130 for how to use your oven for a cozy rising place) or in the bread machine on the programmed day 1 cycle for 18 to 20 hours.

The next morning, if you are using a programmable bread machine, restart the day 1 cycle and allow it to knead for 10 to 13 minutes. If you are making bread using a mixer or by hand, beat the sponge for 10 to 13 minutes until it becomes very smooth. While the sponge is mixing (in the mixer or bread machine), stir the whole grain bread improver into the bread flour. This should be done before adding anything to the bread machine pan or mixing bowl. After the mixing time finishes and the sponge is very smooth, add all the dough ingredients below the line to the bread machine pan, hand mixing bowl, or bowl of your mixer.

If using a bread machine, start the day 2 cycle on the previous page. Add the "adjustment" amounts of water or flour in the ingredient list during kneading if needed to make a supple dough. (Flour is most often needed; water is usually not needed). Spread the dough evenly in the pan at the end of the kneading time to avoid short slices at the ends of the loaves. Allow the cycle to run to completion.

To make the bread with a heavy-duty mixer such as a Kitchen Aid™, after beating the sponge for 10 to 13 minutes on day 2 as directed above, add all of the ingredients below the line and knead the bread according to the mixer directions.

To make the bread by hand, mix the whole grain bread improver into at least half of the bread flour. Add the mixture and the rest of the dough ingredients (reserving about ½ to 1 cup of the flour) to the sponge and mix with a wood or plastic spoon. Stir in as much of the reserved flour as the dough will take. Then knead the dough by hand, using the rest of reserved flour or a little more to knead until the dough is elastic and smooth.

For hand or mixer made bread, form the dough into a loaf and place it in an oiled loaf pan. Let it rise in a cozy place until doubled, about 3 to 3½ hours. Remove the loaf from the oven if that is your cozy rising place. Preheat the oven to 375°F. Bake for 55 to 75 minutes, checking the bread occasionally. Take it out when brown and the loaf sounds hollow when thumped on the bottom.

Cool the bread on a cooling rack completely before slicing it. Makes one loaf of bread, or 15 slices which each contain 1½ carbohydrate units.

One batch of dough for this bread makes about 12 hamburger buns and about 40 rolls. See pages 147 to 149 for directions for making buns and rolls.

TRADITIONAL WHOLE WHEAT FLOUR version of this recipe: Make this bread as described above using traditional King Arthur™ whole wheat flour milled from darker red wheat when you make the sponge on day 1. On day 2, use 2⅛ cups King Arthur™ bread flour with up to 2 tablespoons or more of bread flour for final adjustment of the dough consistency if needed.

White Whole Wheat Sourdough Sandwich Bread

*This recipe was developed to contain only **one** carbohydrate unit per slice so two slices can be used for a sandwich without exceeding the two unit per meal limit for carbohydrates on the insulin resistance diet. It contains much more vital gluten than the previous recipe, so it can seem rubbery at times. However, it makes great grilled cheese sandwiches. The crisp grilled crust outshines "rubbery."*

Ingredients for 1 loaf:	Amount
King Arthur™ white whole wheat flour	1¼ cups
Vital gluten	¾ cup
Florapan™ starter	¼ teaspoon
Non-chlorinated water, 85°F	1½ cups
---	---
King Arthur™ bread flour	1¼ cups + 1-2 tablespoons to adjust
Vital gluten	¾ cup
Sea salt	1¼ teaspoons
Berlin Seeds™ stevia extract	Optional - range ⅛ to ½ teaspoon
SAF™ instant yeast	2 teaspoons
Apple juice concentrate	2 tablespoons
Non-chlorinated water, 85°F	¼ cup

Cycle for day 1 and for mixing on day 2: Dough cycle or this programmable cycle: Knead 1 – 15 minutes; Rise 1 – 23 to 24 hours, Rise 2 and 3 – off, Bake – off.

Cycle for day 2 for bread machine-baked loaf: Knead 1 – 25 minutes, Rise 1 – 1 hour 30 minutes, Rise 2 and 3 – off, Bake – 1 hour.

In the late morning or early afternoon of the day before you want sourdough bread, mix the sponge ingredients above the line in the ingredient list using a wooden or plastic spoon in a glass or ceramic mixing bowl or using a mixer, a bread machine's dough cycle, or the day 1 cycle above. If not using a bread machine, cover the bowl containing the sponge with plastic wrap and a towel. Also cover the pan from a non-programmable bread machine at the end of the dough cycle. Place this pan or the mixing bowl in a cozy (85°F to 90°F) place for the dough to rise.

The sponge should rise for 18 to 20 hours in a cozy place in your kitchen, in your oven (see the second and third paragraphs on page 130 for for how to use your oven for a cozy rising place) or in a programmable bread machine in the "rise" part of the programmed day 1 cycle above.

The next morning, on the second day, if you are using a programmable bread machine, restart the day 1 cycle and allow it to knead for 10 to 13 minutes. If you are making bread using a mixer or by hand, beat the sponge for 10 to 13 minutes until it becomes very smooth. During the mixing time in the bread machine or while your mixer is running, stir together the bread flour and vital gluten. Add the flour-gluten mixture and all of the remaining dough ingredients below the line in the recipe to the sponge.

If using a programmable bread machine, start the day 2 cycle on the previous page. Add the "adjustment" amounts of water or flour during kneading if needed to make a supple dough. (Flour is most often needed; water is usually not needed). Spread the dough evenly in the pan at the end of the kneading time (to avoid having short slices on the ends of the loaves) and allow the cycle to run to completion.

To make the dough with a heavy-duty mixer such as a Kitchen Aid™, while beating the sponge for 10 to 13 minutes as directed above, stir the vital gluten into the bread flour. Then add the flour-gluten mixture and the rest of the dough ingredients below the line and knead the bread according to the mixer directions. Oil a loaf pan whle the mixer is kneading. When the kneading is finished, form the dough into a loaf. Put the loaf into the pan, oil the top of the dough, and cover it with plastic wrap and a towel. Allow the dough to rise in a cozy (85° F) place until it is just under doubled in volume, about 1½ hours. Proceed as in the second paragaph below.

To make the bread by hand, first stir the vital gluten into the bread flour. Then add the flour-gluten mixture and the rest of the dough ingredients below the line to the sponge, reserving about ½ to 1 cup of the flour, Mix with a wooden or plastic spoon, stirring in as much of the reserved flour as the dough will take. Knead the dough by hand, using the remaining reserved flour, or a little more, to knead the dough until it is smooth and elastic. Oil a loaf pan and form the dough into a loaf. Put the loaf into the pan, oil the top of the dough, and cover it with plastic wrap and a towel. Let the dough rise in a cozy (85° F) place until it is just under doubled in volume, about 1½ hours.

After the loaf is finished rising, remove it from the oven (if that is your cozy rising place). Preheat the oven to 375°F. Bake the bread for 55 to 65 minutes, checking the bread occasionally. Remove it from the oven when brown and the loaf sounds hollow when thumped on the bottom.

Cool the bread on a cooling rack completely before slicing it. Makes one loaf of bread, or 14 slices, each of which is one carbohydrate unit, which makes this bread ideal for using two slices for a sandwich.

Buckwheat Sourdough Bread

Sourdough bread has a lower GI score than regular bread made with the same grain. Now, with the new freeze-dried starter, we can make sourdough bread which is wheat and gluten-free and without keeping a starter.

Sponge

2⅓ cups buckwheat flour such as Bob's Red Mill™ brand
¼ teaspoon Florapan™ sourdough starter (See "Sources," page 216).
1⅓ cups purified (non-chlorinated) water

Dough

¾ cup buckwheat flour
2 cups tapioca starch
1½ tablespoons guar or xanthan gum
1⅞ teaspoon sea salt
1½ tablespoons caraway seed (optional)
¾ teaspoon SAF™ instant yeast
3 tablespoons oil
4 to 5 large eggs* at room temperature
1½ tablespoons apple juice concentrate
Enough warm non-chlorinated water to bring the total volume of eggs,
    apple juice and water up to up to 1 cup

To make this bread with a programmable bread machine, program one of the homemade menu cycles on your machine for these times for day 1:

Knead : 10 minutes
Rise 1: 20 to 24 hours.
All the other parts of the cycle should be set to zero minutes.

Set the day 2 cycle to:

Knead: 10 minutes
Rise 1: 2½ hours
Rise 2 and 3: Off
Bake: 60 minutes.

To make this bread without a programmable bread machine, plan to be home for most of two days to keep the sponge and dough at cozy temperature. Your mixer, set on medium speed, does the mixing to develop the guar or xanthan gum structure or the dough cycle of a bread machine can be used to develop the gum structure. Do not keep the sponge or dough in a metal mixing bowl; rather use plastic or glass.

About noon of the day before you want to serve this bread, measure the ingredients for the sponge. Mix them with your mixer for 5 minutes. Cover the glass or plastic mixing bowl with plastic wrap and then with a towel and let it rise in a warm (85 to 90°F) place** for 18 to 20 hours.

If you are using a programmable bread machine, put ingredients for the sponge into the pan of your machine and start the dough cycle or the day 1 programmed cycle. Use a spatula to make sure that all of the ingredients are well combined. Allow the sponge to rise in the machine for 18 to 20 hours on the programmed cycle. If you are using the dough cycle on a non-programmable bread machine, remove the pan from the machine after the dough cycle finishes, cover it with plastic wrap and a towel, and put the pan in a cozy place for 18 or more hours. After this time the sponge should have expanded in volume by one third or more and should smell sour.

On the second day, mix the guar or xanthan gum with the ¾ cup buckwheat flour and tapioca starch. If you are using a programmable bread machine, add the flour mixture and the rest of the dough ingredients below the line to the sponge and start the day 2 programmed cycle. Use a narrow spatula to assist the kneading. For a progammable machine, level the dough in the pan after the kneading. Allow the cycle to run to completion. For a non-programmable machine, transfer the dough to a prepared baking pan, etc. as in the directions in the second paragrah below.

If you are using a mixer, add the flour-gum mixture and the rest of the dough ingredients below the line to the sponge and beat it for 5 minutes. Allow it to rise in your warm place for 30 to 40 minutes and then beat it again for 5 minutes.

While the dough is mixing for the final time, oil and flour a 9 by 5 inch loaf pan. When the dough is finished mixing, scrape it from the mixing bowl into the pan, oil the top of the dough in the pan and cover it with plastic wrap and a towel. Allow the loaf to rise in a warm** (85 to 90°F) place for about 2½ hours or until it barely doubles in volume.

When the loaf has nearly doubled, remove it from the oven (if that is the cozy spot). Preheat the oven to 375°F. Bake the loaf for about 50 to 65 minutes, loosely covering it with foil after the first 30 to 45 minutes if it is getting excessively brown. Remove the loaf from the pan and cool it completely on a cooling rack. The final loaf should sound hollow when tapped on the bottom. The loaf wiil be flat-topped and about 3½ inches tall. Makes one loaf or 15 slices of bread each of which contains 2 units of carbohydrate.

*Note on eggs: If you are allergic to eggs, use warm water in their place. If you do not take the eggs out of the refrigerator early enough for them to come to room temperature before you are ready to bake, put them in a bowl of warm water for 5 or 10 minutes before cracking them to use in this recipe.

**Note on a warm rising place: To use your electric oven as a cozy place for bread to rise, see the second and third paragraphs on page 130. Since your oven will not retain the heat for the 18 to 20 hours needed to ferment the sponge, you will have to adjust the temperature by monitoring the oven temperature with a thermometer and turning the oven on for about 20 to 30 seconds occasionally to raise the temperature as needed. If you have a gas oven, the pilot light may keep it at the right temperature for bread to rise.

100% Whole Wheat Pizza

Pizza is the food that people on restricted diets seem to miss most. Pizza top-pings contain fat and cooked tomatoes so are a good source of lycopene, making this recipe good for eyes are well as a treat. Turn the page for a wheat and gluten-free quinoa pizza recipe.

Ingredients for one whole wheat crust

½ cup water
1½ teaspoons single source honey or agave (to feed the yeast)
2½ teaspoons oil
½ teaspoon salt
1⅓ cups stone ground whole wheat flour such as Bob's Red Mill™ brand
1 tablespoon plus 1 teaspoon vital gluten such as Bob's Red Mill™ brand
1 teaspoon active dry yeast

Toppings

½ batch of pizza sauce, recipe on page 166
6 ounces (1½ cups) grated mozzarella cheese or alternative cheese such as almond cheese, Funny Farm™ goat mozarella, etc.
2 tablespoons Romano cheese (Pecorino Romano is made from sheep milk).
6 ounces cooked lean ground beef, turkey or other cooked meat
Vegetables toppings such as sliced bell peppers, olives, etc. (optional)

Prepare the pizza sauce as directed on page 166. You may wish to double this dough recipe and make two pizzas, or freeze the half-batch of leftover sauce.

Prepare the dough by the mixer method as described in the second and third paragraphs of the bread recipe on page 131 or with the dough cycle of a bread machine. If you use a bread machine, add the ingredients to the pan in the order listed in your bread machine's instruction manual. (For most machines this will be the order given above). Start the dough cycle. After 5 to 10 minutes of mixing, touch the dough. If it is sticky, add a tablespoon of flour. After a few more minutes of mixing, re-check the consistency of the dough and if it is still sticky, add another tablespoon of flour. When it is supple but not sticky, you have added enough flour.

If you live in a humid climate and your flour absorbs moisture from the air, you may need to add a total of 1 to 4 additional tablespoons of flour to the dough to achieve the right consistency – supple but not sticky. In a dry climate during a dry season, you may need to add ½ to 2 teaspoons of water (½ teaspoon at a time) to make the dough supple. Allow the dough cycle to run its course or allow your mixer-made dough to rise in a warm place. As the rising time nears completion, prepare your oven to be a cozy rising place (85 to 90°F) as described on page 130. If you have a gas oven, the pilot light may keep it at the right temperature). Very lightly oil a 12-inch pizza pan by rubbing it with paper towel with just a little oil on it.

When the dough cycle has reached completion or the mixer-made dough has doubled in size, transfer it to the pizza pan. Stretch dough thinly in the pan. Let it rise a cozy place for 5 minutes, top it with the toppings, and then, if needed, let it rise for an additional 5 to 10 minutes or until it is about double its original size.

Remove the pizza from the oven (if that is the cozy rising place) and preheat your oven to 400°F. Bake the pizza for 18 to 20 minutes or until it is nicely browned. Cut the pizza into six slices. Each slice of pizza is one carbohydrate unit. Using the amounts of cheese and meat in the ingredient list, each slice contains 2 protein units. Leftover pizza freezes well.

Quinoa Pizza

This grain and gluten-free recipe is relatively easy to prepare and delicious.

Ingredients for one quinoa crust

⅔ cup water
⅓ cup apple juice concentrate, thawed
2 tablespoons oil
A scant ¼ teaspoon salt
1⅓ cups quinoa flour
1½ teaspoons guar gum
1½ teaspoons active dry yeast

Toppings

½ batch of pizza sauce, recipe on page 166
6 ounces (1½ cups) grated mozzarella cheese or alternative cheese such as
    almond cheese, Funny Farm™ goat mozzarella, etc..
2 tablespoons Romano cheese (Pecorino Romano is made from sheep milk).
6 ounces cooked lean ground beef, turkey or other cooked meat
Vegetables toppings such as sliced bell peppers, olives, etc. (optional)

To make this dough with a mixer, put the quinoa flour, yeast, guar gum, and salt in a large mixer bowl. Mix on low speed for about 30 seconds. In a small saucepan, warm the water, apple juice and oil to 115 to 120°F. (Measure the temperature with a digital or a yeast thermometer). With the mixer running on low speed, add the liquids in a slow stream to the dry ingredients. Continue mixing until the dry and liquid ingredients are thoroughly mixed. Beat the dough on medium speed for 3 minutes. Scrape the dough from the beaters and the sides of the bowl into the bottom of the bowl. Cover the bowl, put it in a warm place (85°F to 90°F; see page 130 for more about how to achieve this) and let the dough rise for one hour or until doubled. Beat the dough again for three minutes at medium speed.

If you have a bread machine, use the dough cycle to make the dough. Mix the guar gum into the flour first. Then add the crust ingredients above to the machine in the order your instruction manual specifies. (For most machines this will be the order given above). Run the dough cycle to completion. Then restart the cycle and allow it to knead for 3 to 5 minutes after it begins kneading quickly if the knead cycle has a gentle start.

Lightly oil a 12-inch pizza pan or baking sheet. Scrape the dough out of the bowl or bread machine pan and onto the prepared pan. Then oil your hand and use it to spread the dough out to the edges of the pizza pan or to flatten it to the desired thickness on a baking sheet.

Preheat your oven to 400°F. While the oven is heating, spread the pizza dough with ½ batch of pizza sauce and the cheese and meat toppings listed or more toppings to provide satiety. Add at least three ounces of protein (meat plus cheese) per serving to make ¼ of a pizza enough to satisfy and to balance the carbohydrate in the crust. Top with vegetable toppings if deisred. Bake for 20 to 25 minutes or until the edge is golden brown. Makes 4 servings, each of which contains 2 carbohydrate units and 3 protein units.

Buns and Rolls

The brief directions for the recipes that follow are based on two assumptions: (1) that "the cook" for most macular degeneration patients is about the age of the patient (probably a senior) and (2) that therefore the cook has seen or done a little yeast bread or roll baking. These assumptions may not be true. Therefore, if you want to make buns, or even just to think about making burger buns or rolls for your loved one, use the contact information at the bottom of page 4 to contact me, and I will send you an e-book of the third edition of *Easy Breadmaking for Special Diets* at no charge to help you get started on yeast bread baking. *Easy Breadmaking* contains recipes of all kinds including "normal" as well as special diet recipes.

Burger Buns

This recipe gives us a wonderful way of making salmon, which can become tiresome when eaten weekly or more often, a treat. Use these buns for "Salmon Burgers," page 100, with all the trimmings. If making gluten-free or sourdough buns, see "Sources," page 211, to purchase a burger bun pan.

> One batch or half-batch of yeast bread dough made with wheat or spelt,
>   pages 131 to 134 and 139 to 142 in this book OR
> One batch or half-batch of gluten-free yeast dough pages 135 or 143
> Monounsaturated oil such as olive, avocado or canola oil

Prepare the yeast dough by hand, mixer, or using the dough cycle of a bread machine as directed in the recipe. After the bread machine dough cycle or first rising time is completed, stir or punch down the dough. If making sourdough, after the dough has risen about two hours on day 2, punch or stir down the dough. Lightly oil burger bun pan(s) or, for non-sourdough wheat or spelt buns, oil a baking sheet or two.

For non-sourdough wheat or spelt dough, divide the dough into pieces that weigh about 3 ounces each. Knead them briefly on a very lightly oiled countertop or board to form them into balls. Place the balls on an oiled baking sheet at least 3 inches apart. Press the balls with your hand to flatten them a little. Oil the tops of the balls and cover the baking sheet with plastic wrap and a towel. Allow the buns to rise in a warm place until doubled in volume, about 45 minutes. Bake at 375°F for 20 to 30 minutes, or until lightly browned.

For sourdough wheat buns, place a 3-ounce ball of dough into each oiled well of the burger pan. Oil the tops of the balls and cover them with plastic wrap and a towel. Allow to rise in a warm place until double in volume, about 1½ hours to 2½ hours. Bake at 375°F for 20 to 30 minutes or until lightly browned.

For gluten-free buns, mix the risen dough for 5 to 10 minutes with your mixer to develop the structure of the guar or xanthan gum. If using a bread machine, re-start the dough cycle and let it mix for 5 to 10 minutes. Lightly oil the wells of burger bun pan(s). Spoon a 3-ounce portion of dough into each well. Oil the top of the dough and cover the pan(s) with plastic wrap and a towel. Allow the buns to rise in a warm place until double in volume, about 40 to 45 minutes for yeast dough or up to two hours for sourdough. Bake at 375°F for 20 to 30 minutes, or until lightly browned.

Each 3 ounce bun contains two carbohydrate units. The number of servings this recipe makes varies depending on how much dough is made. One batch of dough from the "Whole Wheat Sourdough" recipe on page 139 makes about 12 buns. One half-batch of any of the dough recipes listed in the ingredients will make about 5 to 6 buns, just right for one burger bun pan. If you run out of wells in burger pan(s) before you run out of dough, make the leftover dough into the rolls which follow.

Basic Rolls

These are dinner rolls which can be made with any kind of dough. For fancier rolls made with wheat or spelt dough, see the next recipe.

One batch or half-batch of yeast bread dough made with wheat or spelt,
pages 131 to 134, or whole wheat sourddough, page 139 OR
One batch or half-batch of gluten-free yeast dough pages 135 or 143
Monounsaturated oil such as olive, avocado or canola oil

Prepare the yeast dough by hand, mixer, or using the dough cycle of a bread machine as directed in the recipe. After the dough cycle or first rising time is finished, or for sourdough, after the dough has risen for an hour or two on day 2, stir or punch down the dough.

For gluten-containing wheat or spelt dough, divide the dough into pieces that weigh about 1½ ounces each. Knead them briefly on a very lightly oiled board to form them each into a ball. Lightly oil muffin tins. Place a ball into each muffin cup. Oil the tops of the rolls and cover the sheet with plastic wrap and a towel. Allow to rise in a warm place until double in volume, about 40 to 45 minutes for yeast dough or up to two hours or more for sourdough. Bake at 375°F for 20 to 30 minutes, or until lightly browned.

For gluten-free dough, mix the risen dough for 5 to 10 minutes with your mixer to develop the structure of the guar or xanthan gum. If using a bread machine, restart the dough cycle and let it mix for 5 to 10 minutes. Lightly oil muffin tins. Spoon a 1½-ounce portion of dough into each muffin cup. Oil the tops of the rolls and cover them with plastic wrap and a towel. Allow to rise in a warm place until double in volume, about 40 to 45 minutes for yeast dough or up to two hours or more for sourdough. Bake at 375°F for 20 to 30 minutes, or until lightly browned.

Each 1½ ounce roll contains one carbohydrate unit. The number of servings this makes varies depending on how much dough is made. One batch of dough from the "Whole Wheat Sourdough" recipe on page 139 makes about 40 rolls.

Cloverleaf Rolls or Fantans

One batch or half-batch of bread dough made with wheat or spelt,
     pages 131 to 134 and 139 in this book
Monounsaturated oil such as olive, avocado or canola oil

Prepare the yeast dough by hand, mixer, or using the dough cycle of a bread machine as directed in the recipe. After the cycle or first rising time is finished, punch down the dough.

For cloverleaf rolls, divide the wheat or spelt dough into pieces that weigh about ½ ounce each. Knead them briefly on a very lightly oiled board to form them each into a ball. Lightly oil muffin tins. Place about 1½ ounces of balls into each muffin cup. Allow to rise in a warm place until double in volume, about 40 to 45 minutes for yeast dough or up to two hours or more for sourdough. Bake at 375°F for 20 to 30 minutes, or until lightly browned.

For fantans, roll the dough out on an oiled bread board with an oiled rolling pin to ¼-inch thickness or a little less. Cut it into strips 1½ inches wide. Stack 4 to 6 strips, brushing lightly with oil between layers. Cut the stacks of strips into 1½-inch segments, and put a stack of squares of dough weighing about 1½ ounces into each muffin cup with the cut edges of the squares up. Allow to rise until double in volume, about for 45 minutes for yeast dough or up to two hours for sourdough. Bake at 375°F for 20 to 30 minutes, or until lightly browned.

Each 1½ ounce roll contains one carbohydrate unit. If you prefer larger rolls, place more balls or square of dough in each muffin cup. Divide the weight of the larger rolls in ounces by 1½ to caluculate the number of carbohydrate units per roll. The number of rolls this makes varies depending on how much dough is made and the size of the rolls.

Treats

All human beings are born with a desire for sweet foods such as mother's milk. Because treats can be sanity-savers during times of duress, sometimes we still need a sweet or savory treat. Thankfully, we now have healthy sweeteners to use instead of sugar or corn syrup.

This chapter includes recipes for fruit desserts, cookies, puddings and pies. All are sweetened with fruit sweeteners, stevia or monk fruit extract or agave. For savory snacks, see pages 160 to 163. Enjoy these healthy treats!

Blueberry Tapioca

This dessert is great as part of an oven meal and is loaded with anthocyanins for eye health and cinnamon for blood sugar stability.

> 1 16 ounce bag of frozen blueberries
> ¼ cup quick-cooking (granulated) tapioca
> 1 cup apple juice or 1 cup water plus ¼ to ½ teaspoon Berlin Seeds™ stevia
>   extract or pure monk fruit extract, to taste
> ½ to 1 teaspoon cinnamon, to taste

Combine all of the ingredients in a 1½-quart casserole dish. Bake at 350°F for 40 to 60 minutes or until the tapioca is clear. If sweetend with stevia or monk fruit, makes 4 servings each containing one carbohydrate unit. Makes 5 servings each containing one carbohydrate unit if made with the apple juice.

Cherry or Berry-Banana Sorbet

This is simple to make from frozen berries without needing to freeze bananas in advance. It is also full of eye-supportive anthocyanins from the dark fruit.

> 2 ripe bananas
> 2 tablespoons agave or ⅛ to ¼ teaspoon Berlin Seeds™ stevia extract or
>   pure monk fruit extract (optional, to taste)
> 4 to 6 cups frozen dark (Bing) cherries, blueberries, or blackberries

Puree the bananas and optional sweetener in a food processor or blender or blend with a hand blender until smooth. Gradually add the cherries or berries, processing after each addition, until the sorbet reaches the desired consistency. Serve immediately. Freeze any leftovers. Remove the leftovers from the freezer about 20 minutes before serving. Makes about 4 cups of sorbet, or 5 servings each containing 2 carbohydrate units.

Baked Apples or Pears

This homey dessert is wonderful as part of an oven meal. The cinnamon pro-
motes stable blood sugar levels.

> 4 large baking apples or pears*
> ½ cup apple juice, pear juice, or water OR ⅛ teaspoon Berlin Seeds™
>   stevia extract or pure monk fruit extract + ½ cup water
> 1 teaspoon cinnamon
> ¼ cup raisins, goji berries or other eye-supportive dried berries (optional,
>   see "Sources," page 212 for more dried berry options)

Core the apples or pears and put them in a 2½ quart glass casserole dish with a lid. Pour in the juice, water, or water plus sweetener extract and sprinkle the cinnamon down the centers of the fruit. Stuff the fruit with raisins or berries, if desired. Bake at 350°F for 40 to 50 minutes for the apples or 1 to 1½ hours for the pears or until the fruit is tender when pierced with a fork. Makes 4 servings.

*Note on apples and pears for baking: Some varieties of apples hold their shape well when they are baked, such as Rome, Pink Lady, and Granny Smith. If this dessert is for guests and you will be buying the apples especially for baking, choose a baking apple. However, if there are apples in your refrigerator, any variety will work in this recipe. I personally think Bosc pears are the most delicious for baking because they are thin skinned and sweet, but any kind of pears will work in this recipe.

Cookies

High-Protein Brownies

> 8 ounces of natural nut butter such as almond butter
> 2 large eggs
> ¼ cup plus 1 tablespoon water
> 1 teaspoon vanilla (optional)
> ⅛ teaspoon salt
> ⅞ teaspoon Berlin Seeds™ stevia extract or pure monk fruit extract
> ¼ cup plus 1 tablespoon cocoa (natural, not Dutch process or alkalinized)
>   or cacao powder
> ¼ teaspoon baking soda

Preheat your oven to 350°F. Line the bottom and an inch or two up the sides of an 8 or 9 inch square baking pan with parchment paper.

Measure the cocoa or cacao powder into a small bowl. Put the nut butter in a food processor bowl with the metal pureeing blade. Add the eggs, water, vanilla, salt and stevia or monk fruit extract and process until well mixed, stopping and scraping the bowl a time or two. Stir the baking soda into the cocoa in the small bowl. Add it to the food processor bowl and very briefly mix it in with a spatula until there is only a little dry powder on the top of the mixture. Process for a few seconds until just blended. Scrape the batter into the prepared pan and bake for 20 minutes. Cool completely before cutting. Makes 12 to 16 brownies which each contain zero carbohydrate units.

Carrot Cookies

These cookies supply vitamin A and can be sweetened with stevia, monk fruit, or fruit juice.

> 3 cups quinoa or spelt flour
> 1 cup tapioca starch or arrowroot
> 1½ teaspoons baking soda
> ⅜ teaspoon unbuffered vitamin C crystals if using the apple juice OR
> > ½ teaspoon unbuffered vitamin C crystals if using the stevia or monk fruit extract and water
> 1½ teaspoons cinnamon
> 2¼ cups grated carrots
> 1 cup raisins or chopped dates or other dried fruit (optional)
> Choose one sweetener/liquid
> > 1⅜ cups apple juice concentrate, thawed,
> > 1⅜ cups water plus ¾ to 1 teaspoon Berlin Seeds™ stevia extract or
> > > pure monk fruit extract
> ½ cup oil

Preheat your oven to 350°F. Mix together the flour, starch, baking soda, vitamin C crystals, cinnamon and stevia or monk fruit extract (if using them) in a large bowl. Stir in the carrots and raisins or chopped dried fruit. Combine the juice or water and oil and stir them into the flour mixture until they are just mixed in. Drop the batter by heaping teaspoonfuls onto an ungreased cookie sheet and bake for 12 to 15 minutes. The stevia or monk fruit sweetened cookies will not brown much, but they will feel dry when touched. Remove them from the cookie sheets and place them on paper towels to cool. Makes 4 to 5 dozen cookies each of which contains 2/3 carbohydrate unit if made with apple juice or ½ carbohydrate unit if made with stevia or monk fruit. The raisins add about 1 gram of carbohydrate per cookie, a negligible amount if the serving size is not extremely large.

No-Carb Cacao, Chocolate or Carob Chip Cookies

3½ cups finely ground blanched almond flour such as Honeyville™ brand
¼ teaspoon salt
½ teaspoon baking soda
¾ teaspoon Berlin Seeds™ white stevia powder OR 1 teaspoon monk fruit
    extract powder, or to taste
2 to 3 eggs plus enough water to bring their volume to ½ cup
½ cup oil
1 teaspoon vanilla extract (optional)
½ cup cacao or carob chips, recipes on pages 170 to 171, or purchased
    chocolate or carob chips ("Sources," p. 219) that meet your health needs

Line two large baking sheets with parchment paper. Preheat your oven to 350°F. Mix together the flour, salt, baking soda and stevia or monk fruit powder in a large bowl. Break the eggs into a measuring cup and beat lightly with a fork. Add water to bring the volume in the cup up to ½ cup. Then add the oil and vanilla to the cup and stir. Before the liquids can separate, add the liquid ingredients to the flour mixture and stir until they are mixed in. Stir in the chips. Drop the batter by level tablespoonfuls onto the prepared cookie sheets, leaving at least 2 inches between the cookies. Flatten the dough with your fingers held together. Bake for 7 to 10 minutes or until the cookies are brown on the edges. Allow them to stand on the baking sheets for at least 20 minutes after removing them from the oven. Then slide the parchment paper and cookies onto a table or countertop to finish cooling. Makes 28 to 30 cookies each of which contains less than one gram carbohydrate (zero carbohydrate units) and ¾ protein unit if it contains carob or cacao chips made from the recipes on pages 170 to 171. If made with purchased carob or chocolate chips, each cookie may contain (depending on the chips used) 2/3 carbohydrate unit and ¾ protein unit.

Puddings

Blueberry Pudding

1 pound of frozen blueberries, thawed
¾ cup water with large blueberries*
1 teaspoon Berlin Seeds™ stevia extract or pure monk fruit extract
⅛ teaspoon vanilla
1 tablespoon psyllium (Indian husk) powder

Puree any juice in the fruit bag with a few blueberries. Add the water, stevia or monk fruit and vanilla and blend. Working quickly, stir in the Indian husk powder thoroughly. Still working quickly, stir in the blueberries. Put the pudding in a bowl and refrigerate, or while stirring let it thicken slightly more, and put it in a pie crust and refrigerate the pie. Makes 6 servings each containing 1 carbohydrate unit.

*Note: If using small wild blueberries, increase the water to 1 cup and and 1 addtional teaspoon of psyllium powder

Ricotta Pudding

Whipped cream ingredients
    1 cup (one ½ pint carton) whipping cream
    1½ teaspoons Signature Secrets™ culinary thickener
    ⅛ teaspoon Berlin Seeds™ stevia extract or pure monk fruit extract

Rest of the ingredients;
    1⅔ 15-ounce containers of good quality ricotta such as Galbani™ ricotta
    ⅜ cup milk
    2 to 3 teaspoons vanilla
    1½ teaspoons Berlin Seeds™ stevia extract or pure monk fruit extract

Pour the whipping cream into a food processor with the blade for pureeing. Puree it until it becomes quite thick. Add the thickener and ⅛ teaspoon stevia and puree it until the whipped cream is stiff. Scoop it out of the processor and put it into a bowl, trying to remove most of it with a spatula.

Place the ricotta, milk, vanilla and 1½ teaspoons stevia in the processor and puree. Add the set-aside whipped cream and pulse a few times until it is just mixed. Refrigerate the pudding. Makes six 6-ounce servings which each contain zero carbohydrate units.

Pumpkin Pudding

This pudding contains plenty of vitamin A and carotenoids from the pumpkin.

    1 29-ounce can of thick canned pumpkin such as Kuner's™ brand
    2 teaspoons cinnamon
    1 teaspoon nutmeg
    ½ teaspoon allspice
    2¼ teaspoons Berlin Seeds™ stevia extract or pure monk fruit extract
    6 large eggs
    1¾ cups milk

Whisk or beat the eggs and combine them with the other ingredients in a saucepan. Preheat the oven to 375°F. Cook the pudding over medium heat until steamy. Pour it into one or more casserole dishes, leaving at least an inch or two at the top of the dish for expansion during baking. Bake for 45 minutes or until it is set in the middle and a knife inserted in the middle comes out wet but without pudding on it. This recipe contains zero carbohydrate units.

Pie Crusts and Pies

Almond Pie Crust

This pie crust is a good no-carbohydrate choice for all the pies in this book.

1¾ cups finely ground blanched almond flour
¼ teaspoon unrefined salt
¹/₁₆ teaspoon Berlin Seeds™ stevia extract or pure monk fruit extract
    (Use of a sweetener is optional. Omit or substitute 1 tablespoon agave
    for 1 tablespoon of the water below instead of using the extracts).
¼ cup oil
1 teaspoon vanilla (optional)
2 tablespoons water OR 1 tablespoon water + 1 tablespoon agave

Preheat the oven to 350°F. Stir together the almond flour, salt and optional stevia or monk fruit extract in a large bowl. Add the oil and stir it in thoroughly until it is all taken up by the flour and the dough forms lumps. Add the water or water plus optional agave and/or vanilla and stir until the pie crust comes together. Press the dough on to the bottom and sides of a 9-inch glass pie dish. Bake for 9 to 15 minutes or until the crust is golden brown. Cool the crust.

If you will be filling this with a "wet" filling such as pumpkin (see page 158), the crust may become soggy if the pie is stored in the refrigerator for a long time and will need the protection of an egg wash.[1] Therefore, again preheat your oven to 350°F. After the crust has cooled enough to handle the pie dish comfortably, slightly beat an egg and brush it on the bottom and sides of the crust. Wrap a 4-inch wide strip of aluminum foil around the edge of the pie dish and fold it over the edge of the crust to protect it from over-browning. Bake for 6 to 9 minutes until the egg is set and dry. Leave the foil on the edge of the crust. Cool the crust completely before filling. Makes one pie crust or six servings each containing zero carbohydrate units.

Coconut Pie Crust

2 cups unsweetened shredded or very finely shredded coconut
Coconut oil – ¼ cup with shredded coconut or ⅜ cup with finely shredded
    coconut

Preheat your oven to 300°F. Melt about ½ cup coconut oil in the microwave or in a small saucepan over low heat. Measure ¼ cup of the liquid coconut oil.

---

1 If you are allergic to eggs, omit the egg wash. A slightly soggy crust is better than an allergic reaction. Using high-quality thick pumpkin also helps prevent a soggy crust,.

Measure the coconut into a glass pie dish. Pour the ¼ cup coconut oil over the coconut. Mix the oil and coconut thoroughly using a spoon and your hands. If necessary, add more melted coconut oil to get the coconut to stick together. (More oil will be needed if the coconut is finely shredded). Press the mixture evenly onto the bottom and sides of the dish.

Bake the crust for 12 to 15 minutes or until it begins to brown. Cool the crust completely on a wire cooling rack. Fill the crust with blueberry or other fruit filling[2] which is thickening but not set. (See page 158 for blueberry pie). Makes one pie crust or six servings each containing zero carbohydrate units.

Quinoa Pie Crust

2 cups quinoa flour
1 teaspoon baking soda
¼ teaspoon unbuffered vitamin C powder
½ teaspoon salt (optional)
½ teaspoon cinnamon (optional)
½ cup oil
4 to 6 tablespoons cold water

Preheat your oven to 350°F. Stir together the flour, salt, baking soda, vitamin C and cinnamon in a large bowl. Add the oil and blend it in thoroughly with a pastry cutter. Add 4 tablespoons of water and mix the dough until it begins to stick together, adding more water if necessary.

Divide the dough in half and press each half of the dough into a glass pie dish. If you are making a type of pie that directs that the crust be baked before adding the filling to it, gently prick the crusts with a fork. Bake the crusts for 20 to 25 minutes or until they begin to brown on the bottom. Completely cool the crusts on a wire cooling rack before filling them. If you are making only one pie, freeze the second crust.

Although this crust not as likely to become soggy as an almond crust, it may be coated with an egg wash[3] if used for pumpkin and other liquid-filling pies. Bake the crust, shielding the edge with a strip of foil wrapped around the pie dish and folded over the edge of the crust. Cool the crust, brush it with beaten egg, and bake it again until the egg is set. Cool the crust again and add the pumpkin filling and bake as directed in the recipe on page 158. Remove the foil from the crust's edge for the last ten minutes of baking. Makes two pie crusts containing six servings each. Each sixth-of-the-pie serving of crust contains 1½ carbohydrate units.

---

2 Use the contact information at the bottom of page 4 to contact me for other fruit filling recipes (cherry, apple, etc.) which work well with this crust.

3 If you are allergic to eggs, omit the egg wash. A slightly soggy crust is better than an allergic reaction.

Quick-Mix No-Roll Amaranth Pie Crust

This amaranth crust has a delicious nutty taste and is easy to make.

1½ cups amaranth flour
¾ cup arrowroot
½ teaspoon salt
½ teaspoon baking powder
⅝ cup oil
¼ cup cold water

In a large bowl, stir together the flour, arrowroot, salt and baking powder. Measure the water and oil into the same measuring cup and stir them together thoroughly. Before they have a chance to separate, pour them into the bowl with the flour. Stir the ingredients together quickly; if the flour does not all incorporate into the dough readily, cut the mixture with the side of the spoon to help the dough come together into a crumbly mixture. Do not stir or cut the dough for very long; do not over-work the dough.

For two single pie crusts, press half of the dough on to the bottom and sides of each of two pie plates. Preheat your oven to 400°F. Bake for 10 to 15 minutes or until the pie crust is lightly browned.

Makes two single pie crusts or 6 servings per crust each containing ¾ carbohydrate unit.

Stone Ground Whole Wheat Pie Crust

1¼ cups stone ground whole wheat flour
¼ to ½ teaspoon salt
¼ teaspoon baking soda (or use baking powder and omit the vitamin C)
1/16 teaspoon unbuffered vitamin C powder or crystals
½ cup Spectrum Naturals™ shortening
4 to 7 tablespoons very cold water

Preheat your oven to 400°F. In a large bowl, stir together the flour, salt, baking soda and vitamin C. Cut in the shortening until the consistency of corn meal. A few pea-sized pieces of shortening may remain, and this is fine. Add 4 tablespoons of the water and stir. If the dough is not coming together, add another 2 to 3 tablespoons of water and mix until the dough begins to come together. Press the dough into a ball with your hands. Place the dough on a floured pasty cloth. Roll with a floured rolling pin to about an eighth inch in thickness. Use the pastry cloth to gently fold the crust in half. Transfer the dough to a glass pie plate and unfold it. Trim and crimp the edge of the crust. Prick the bottom and sides of the crust with a fork. Bake for 9 to 11 minutes or until it begins to brown. Makes one pie crust or six servings each containing 1½ carbohydrate units.

Pumpkin Pie

Filling ingredients
>1 15-ounce can of thick canned pumpkin such as Kuner's™ brand
>1 teaspoon cinnamon
>½ teaspoon nutmeg
>¼ teaspoon allspice
>1 to 1⅛ teaspoon Berlin Seeds™ stevia extract or pure monk fruit extract.
>3 large eggs
>⅞ cup milk

1 single Almond Pie Crust, page 155, whole wheat crust, page 157, or other crust, baked and cooled
1 egg for brushing the crust if not allergic to eggs[4]

Make the crust as directed in the recipe. Cover the edges of the crust with foil. Bake as directed in the recipe until it is set and just beginning to brown in a place or two. Remove it from the oven and let it cool. Brush it with beaten egg and bake for another 6 to 10 minutes or until the egg is set. Cool the crust completely. Leave the foil on the crust edges. Fill and bake as below.

For the filling, whisk or beat the eggs and combine them with the other ingredients in a saucepan. Preheat the oven to 375°F while cooking the filling. Cook the filling over medium heat until steamy. Put the filling into the crust. Bake for 30 to 35 minutes or until the filling is just beginning to set in the middle of the pie. Take the foil off the edges of the crust and bake for another 10 to 15 minutes until the edge is brown and a knife inserted in the filling comes out wet but without filling on it.

Makes one pie or six servings each of which contain zero carbohydrate units if made with the almond crust. If made with any other type of crust, each serving contains the number of carbohydrate units that are in the crust for one serving.

Blueberry Pie

Any single-crust pie crust in this book, baked
One batch of "Blueberry Pudding," page 153

Make the pie crust and allow it to cool. Make the blueberry pudding, stirring it every few minutes as it thickens. When it is very thick, transfer it from the bowl to the crust. Refrigerate the pie. Makes one pie or six servings each of which contain ¾ carbohydrate unit for the filling plus the number of carbohydrate units per serving of the crust.

---

4 If you are allergic to eggs, omit the egg wash. A slightly soggy crust is better than an allergic reaction. Using high-quality thick pumpkin also helps prevent a soggy crust,.

Sugar-Free Whipped Cream

This is delicious on the pies and puddings in this chapter. It is good made without a thickener if you are allergic to corn or gelatin or if you will be making it shortly before serving time.

½ pint (1 cup) whipping cream
⅛ teaspoon of Berlin Seeds™ stevia extract or pure monk fruit extract
 OR 2 teaspoons to 1 tablespoon agave
Optional thickener (Chose one or omit entirely):
 2 teaspoons Signature Secrets Culinary Thickener™ (Omit if corn
 allergic. See "Sources," page 217 to order it).
 1 teaspoon unflavored gelatin PLUS 4 teaspoons water

For natural whipped cream, an hour or two before you plan to serve it, beat the cream with a mixer on medium speed until soft peaks form. Add the sweetener and whip on high speed until stiff peaks form. The cream will have a smooth surface but form stiff peaks when the beater is lifted. It will remain whipped for several hours.

To thicken the whipped cream with Signature Secrets Culinary Thickener™, beat the cream at high speed until soft peaks form. Add the sweetener and beat it in thoroughly. Add the Signature Secrets™ thickener and beat it on low speed until it is thoroughly mixed into the cream. Then beat on high speed for one to two more minutes until the cream forms stiff shapes. The surface of the cream will look chunky if this thickener is used. It will stay whipped a week or more.

To thicken the whipped cream with gelatin, sprinkle the gelatin over the water in a small glass bowl. Allow it to stand for at least 5 minutes. Then microwave it for about 10 seconds, stopping after each 5 seconds and whisking, until the gelatin is completely dissolved. Set aside to cool slightly while you whip the cream and sweetener on medium speed until the beaters begin to leave a track in the cream. Add the gelatin mixture to the cream slowly with the beaters running. Then increase the speed to high and whip the cream until the peaks are stiff. The whipped cream will remain whipped for several days or more.

Store the whipped cream in the refrigerator.

Makes about 2 cups of whipped cream which contains zero carbohydrate units if made with the stevia or monk fruit extract. If made with agave, a whole batch contains one carbohydrate unit, a negligible amount of carbohydrate per serving.

This 'n That

This chapter contains an assortment of recipes of many kinds. Help for when hunger hits between meals is addressed first with advice for healthy snacking and snack recipes. Next, explore the recipes for sauces, dressings and condiments to add zest and enjoyment to meals.

The end of the chapter contains recipes for ingredients required in other recipe chapters. The stevia and monk fruit working solutions are used in "Sugar-Free Cola" and other recipes but can also be used to sweeten coffee and tea. Take a small dropper bottle of these solutions with you when you might be having a hot beverage away from home. This will spare your eyes and blood sugar level from harm done by artificial sweeteners or sugar.

Finally, there are recipes for carob and cacao chips sweetened with stevia or monk fruit extract that are used in cookies and gorp. Carob and cacao candy are larger versions of the chips. The last recipe in this book, "Cacao Candy," is a nutrtional powerhouse and may be nearly as good for the psyche as the first recipe in this book, the "Eye Smoothie," is for eye health.

Enjoy!

Snacks

Snacks are an important part of a healthy eating plan for eyes because they are necessary to control blood sugar levels. If you allow yourself to get to the point of feeling starved, your body responds by mobilizing glucose stored as glycogen in the liver which can set off blood sugar fluctuations. Therefore, do neglect eating healthy protein-containing snacks which are needed to stabilize blood sugar. This chapter includes a list of quick and easy snack ideas, below, plus a few recipes for snacks.

EASY SNACK IDEAS

(1) Nuts or seeds, alone or eaten with fruit. To get one protein unit, have a handful of nuts or seeds or about 1¼ ounces of sunflower seeds, 1 ounce of peanuts or pumpkin seeds, 1½ ounces of cashews, almonds, shelled pistachios or walnuts, or 2 to 3 ounces (two handfuls) of pecans or macadamia nuts.

(2) Cheese. String cheese comes pre-packaged in the right size (one ounce) for a single serving, and one-ounce single-serving cheddar cheese sticks are also sold in large grocery stores.

(3) GORP is "Good Old Raisins and Peanuts" or mixtures of other nuts and raisins or other dried fruits. Mix the amounts of nuts given above (idea 1) with two tablespoons of raisins or other small or cut-up fruit dried without sugar. See the recipe for "Goji GORP" on the next page.

(4) Vegetables or fruit with nut butter or cheese. The only limit on combinations is your imagination. Here are some examples:

An apple with one or more ounces of cheddar cheese or string cheese

A small banana sliced lengthwise and spread with two to three tablespoons of peanut, cashew, or almond butter

A small cantaloupe wedge (¼ of a 5 to 6-inch cantaloupe) or peach half filled with ¼ to ½ cup of cottage cheese

Celery with 2 to 3 tablespoons of natural nut butter spread down the middle of the stalk. Place raisins on top of the nut butter for "bumps on a log."

(5) One carbohydrate unit of crackers (15 grams of carbohydrate) balanced with two tablespoons of natural nut butter or one ounce of cheese. Suggestions for commercially made crackers which are higher in fiber than most include Mary's Gone Crackers™ gluten-free crackers, Wasa™ 100% whole rye crackers, and Ak-Mak™ stone ground whole wheat crackers. (See "Sources," page 219, for where to purchase crackers). A recipe for "Quinoa Sesame Seed Crackers" is on page 128.

Goji GORP

Dr. Lylas Mogk[1] recommends wolfberries, another name for goji berries, for eye health. Here's a delicious way to eat them. Other berries may be used such as eye-healthuy bilberries. The Brazil nuts provide selenium, a vital nutrient for macular degeneration patients.

¾ cup almonds

¾ cup walnuts or pecans

½ cup Brazil nuts, each cut into 3 or 4 pieces if desired

1¼ cups goji berries, bilberries, or other small or cut-up dried fruit (See "Sources," page 212 to purchase bilberries and other dried wild berries)

1 cup monk fruit or stevia-sweetened cacao, chocolate or carob chips, recipes on pages 170 to 171 (optional)

Mix all of the ingredients together in a storage container. Take a bag of this mixture along with you for a healthy energizing snack away from home. Makes about 3½ to 4½ cups of snack mix or 6 servings each containing 1 carbohydrate and 1½ protein units.

---

1 Mogk, Lylas G. MD and Marja Mogk. *Macular Degeneration: The Complete Guide to Saving and Maximizing Your Sight.* (New York, NY: Ballantine Publishing Company, 1999, 2003).

Kale Chips

1 bunch of red or green kale weighing ½ to ¾ pound
½ cup cashew butter
½ cup cooked or canned garbanzo beans, drained
½ cup black olives, drained
¼ cup olive oil
1 tablespoon wine vinegar
¼ teaspoon salt
1 to 3 tablespoons water

Thoroughly wash the kale and let it drain a few minutes. Then blot it on both sides with paper towel and leave it to dry on dry paper towel for an hour or more, but not long enough to wilt. Begin preheating your oven to 350°F when you start making the nut butter mixture.

Combine the garbanzo beans, olives, olive oil and vinegar in a blender or food processor and puree until relatively smooth. Add the cashew butter and puree. Add 1 tablespoon of the water and salt and puree until totally smooth. Add more water to make the mixture thick but not gooey.

Remove the large main stem in each leaf of kale. Cut the leaves into bite-sized pieces, Put about half of the leaves in a large bowl. Add about one-third of the nut mixture. Mix and massage with your hand(s) until the leaves are all lightly coated at least on one side. Add more of the mixture, up to half of the batch, to coat the leaves thoroughly on both sides.

Spread the leaves out in a single layer on a large baking sheet. Bake for 18 to 20 minutes or until crisp. Cool completely on the baking sheet before storing them. Add the nut mixture to the second half of the leaves, mix and massage to coat, and bake them also. Makes one 2-quart tin of chips. Store the chips in a metal tin and add a desiccant pack (that comes in supplement bottles) if available. This snack contains zero carbohydrate units.

Nourishing Nuts

If you have trouble digesting nuts, try this recipe to rid nuts of their natural phytic acid and enzyme inhibitors which make nuts difficult for us to digest.

About 1 pound of raw nuts
1 to 2 tablespoons unrefined sea salt
Purified water

Put the nuts in a bowl or two quart glass canning jars, filling jars to about one inch from the top. Add the salt to the bowl or jars. Fill the bowl to cover the nuts or fill the jars to the top with water and stir or invert a few times to dissolve the salt. Soak the nuts for up to 24 hours. This long soak makes nuts easier to digest. I soak pecans, walnuts, almonds, macadamia nuts, filberts, pine nuts, and Brazil nuts for 24 hours. Cashews only require a 6 hour soak because they are not actually raw nuts. The GAPS[2] diet instructs that nuts be soaked for 24 hours.

Drain the water from the nuts. Spread the nuts in a single layer on dehydrator trays or baking sheets. Place them in a food dehydrator set at 105°F or oven rigged with the door open to maintain a temperature of up to 150°F to dry. Dry until the nuts are dry and crisp. (Let them cool a bit before judging crispness). I dry most nuts for 24 hours. When drying whole macadamias, some of the nuts may remain soft after 24 hours, so I have dried them up to 34 hours. The solution to this problem is to use macadamia pieces which are smaller. See "Sources," page 219, for where to get macadamia pieces. If you are not zealous about raw foods, you may dry macadamia nuts at 145°F to insure complete dryness so that the nuts will not become moldy.

Store the dried nuts in glass jars with the lids a little loose or in cellophane bags. This snack contains zero carbohydrate units.

Nourishing Seeds

About 1 pound of raw seeds such as sunflower or pumpkin seeds
2 tablespoons unrefined sea salt
Purified water

Put the raw seeds in a bowl or two quart glass canning jars, filling jars to about one inch from the top. Add the salt to the bowl or jars. Fill the bowl to cover the seeds or the jars to the top with purified water and stir or invert is a few times to dissolve the salt. Soak the seeds for 8 to 14 hours. Being smaller, seeds take less time to soak than nuts and also dry more quickly.

Drain the water from the seeds. Spread the seeds in a single layer on dehydrator trays or baking sheets. Place them in a food dehydrator set at 105°F or oven rigged with the door open to maintain a temperature of up to 150°F to dry. Dry until the seeds are dry and crisp. Pumpkin seeds may begin to brown when they should be removed from the dehydrator or oven. Smaller pumpkin seeds may be dry in 6 to 8 hours. Sunflower and larger pumpkin seeds and sunflower seeds take 12 to 15 hours.

Store the dried seeds in glass jars with the lids a little loose or in cellophane bags.This snack contains zero carbohydrate units.

---

2 The GAPS diet is a diet for digestive problems which was developed for autistic children. Campbell-McBride, Natasha, MMedSci. *Gut and Psychology Syndrome.* Medinform Publishing, Soham, Cambridge, UK, 2010.

Sauces and Dressings

Mayonnaise

Here's real mayonnaise made with lutein-rich eggs from naturally raised chickens. It is made without vinegar for those who are allergic to yeast. Home-made mayonnaise is so delicious that it's worth the time it takes to make it.

> 1 egg from a naturally raised chicken or ¼ cup pasteurized[3] egg substitute such as EggBeaters™ (if you tolerate the ingredients)
> 1 teaspoon dry ground mustard
> 1 teaspoon salt
> Dash of pepper
> 1 teaspoon agave or $^1/_{16}$ teaspoon Berlin Seeds™ stevia extract or pure monk fruit extract
> 1 cup monounsaturated oil, divided
> 3 tablespoons lemon juice

Combine the egg or egg substitute, mustard, salt, pepper, sweetener, and ¼ cup of the oil in the bowl of a food processor or blender. Turn on the food processor or blender. After the ingredients are thoroughly mixed (this takes just a few seconds), very slowly, pouring in a trickle, add half of the remaining oil while processing continuously. At this point, stop processing. Add the lemon juice and begin processing again. After the lemon juice is mixed in (again, this takes just a few seconds), very slowly, pouring in a trickle, add the rest of the oil while processing continuously. Makes about 1½ cups of mayonnaise. Store in the refrigerator. This recipe contains zero carbohydrate units.

Super-Smooth Sauce

This mayonnaise substitute made from natural nut butters is a good replacement for mayonnaise for those who are allergic to eggs. If you are allergic to lemon or lime juice, use the vitamin C.

> ¼ cup cashew, almond or macadamia nut butter
> ¼ cup lemon or lime juice or ¼ cup water plus 1½ teaspoons unbuffered vitamin C powder
> ⅛ teaspoon salt
> ¼ cup monounsaturated oil

Combine the nut butter, lemon or lime juice or water plus vitamin C, and salt in a blender or food processor, or use a 2-cup glass measuring cup and a hand blender. Blend these ingredients until they are thoroughly mixed. With the blender or pro-

---

3 Young children, the elderly, and others with compromised health should avoid eating raw eggs due to the potential risk of food-borne illness.

cessor running, add the oil very gradually in a thin steam until it has all been added and the sauce is thick, smooth, and creamy. Makes about ¾ cup of sauce. Store in the refrigerator. This recipe contains zero carbohydrate units.

Pesto

Although "pesto alla Genovese" is traditionally made with sweet basil, pine nuts and olive oil, the combinations below are all delicious. Parsley and spinach pesto will provide your eyes with lutein and zeaxanthin.

TRADITIONAL PESTO
½ pound (about 7 cups) sweet basil
1 to 2 cups raw or soaked and dried pine nuts or nuts of your choice
1 to 3 cloves of garlic, optional
1 cup olive oil
1 to 1½ teaspoons salt
¼ teaspoon pepper, optional

PARSLEY PESTO
½ pound (about 7 cups) Italian flat-leaf parsley
1 to 2 cups raw or soaked and dried walnuts or nuts of your choice
1 to 3 cloves of garlic (optional)
1 cup walnut oil (same as the nuts) or monounsaturated oil such as olive oil
1 to 1½ teaspoons salt
¼ teaspoon pepper, optional

SPINACH PESTO
½ pound (about 7 cups) spinach
1 to 2 cups raw or soaked and dried almonds or nuts of your choice
1 to 3 cloves of garlic, optional
1 cup almond oil (same as the nuts) or monounsaturated oil such as olive oil
1 to 1½ teaspoons salt
¼ teaspoon pepper, optional

Remove the stems from the vegetables and wash the leaves in several changes of water. Spread them out on paper towel. Blot them to remove most of the water. If you are using the garlic, chop it in a blender or a food processor with the metal blade using a pulsing action. Add the vegetable leaves to the processor or blender and use a pulsing action to chop the leaves. Add the nuts to the processor or blender and process continually until they are ground. With the machine running, add the oil in a thin stream. Add the seasonings and process briefly. Use pesto over pasta, instead of tomato sauce on pizza, on spaghetti squash, as a condiment, or for a spread on bread or crackers. Makes about 2¼ cups of pesto, or enough for 3 to 4 pounds of pasta. Leftover pesto freezes well. This recipe contains zero carbohydrate units.

Pizza Sauce

This quick and easy sauce is a good source of lycopene. To always be ready for a pizza, make double or larger batches and freeze it for future use.

> 1 6 ounce can tomato paste
> 1 8 ounce can tomato sauce
> ⅓ cup water
> 1 teaspoon dried oregano
> ½ teaspoon dried thyme
> ½ teaspoon dried sweet basil
> 1 tablespoon olive oil (optional but good for absorption of lycopene)

Combine all the sauce ingredients in a saucepan and heat on medium heat until the sauce just comes to a boil. Reduce the heat to low and simmer the sauce for 30 to 45 minutes or until it thickens a little, stirring every ten minutes to keep the sauce from sticking to the bottom of the pan. This makes enough sauce for two pizzas. You can either freeze half of the sauce for future use or double the amounts of dough and toppings you use and have two pizzas. If you do not eat both pizzas, the leftover pizza will freeze well. See pages 145 and 146 for pizza recipes. This recipe contains zero carbohydrate units.

Oil and "Vinegar" Dressing

Because they are not heated, salad dressings are a great place to add fragile oils (canola and walnut) which are high in essential fatty acids to your diet.

> ½ cup oil, walnut or canola because they are high in omega-3 fatty acids, or monounsaturated olive or avocado oil
> ½ teaspoon dried oregano or sweet basil OR 1 teaspoon of a mixture such as Penzeys™ Italian herb mix (optional)
> ½ teaspoon salt
> Dash of pepper
> ⅓ cup wine vinegar, apple cider vinegar, or lemon juice OR 1 to 1½ teaspoons unbuffered vitamin C powder mixed into 2 tablespoons water (See "Sources," page 217 for the vitamin C).

Combine all of the ingredients in a jar and shake. The dressing may be refrigerated at this point. Shake the dressing again to thoroughly mix it right before pouring it on salad. Makes about ¾ cup of dressing which contains zero carbohydrate units.

Coleslaw Dressing

½ cup tahini or light colored smooth nut butter such as blanched almond or
    macadamia butter

2 to 4 teaspoons unbuffered vitamin C powder, to your taste, (see "Sources,
    page 217) or ⅓ cup lemon juice

½ to ¾ teaspoon salt

⅛ teaspoon Berlin Seeds™ stevia extract or pure monk fruit extract OR
    2 teaspoons agave (optional)

Water – ⅔ cup if using the vitamin C, ⅓ cup with the lemon juice

1 cup monounsaturated oil such as olive, avocado or canola oil

1 teaspoon caraway or celery seeds (optional)

Combine the tahini or nut butter, vitamin C powder or lemon juice, salt, sweetener and water in the bowl of a food processor or blender or the container for a hand blender. Blend until smooth. With the blender running, drizzle in the oil in a slow stream. Stir in the seeds. Serve soon or refrigerate any leftover sauce. Makes about 2¼ cups of dressing which contains zero carbohydrate units. Toss with shredded cabbage or jicama to make coleslaw.

Caesar Salad

This salad is impressive enough for guests yet gives us omega-3 fats from a small serving of cold water fish. Be sure to use mayonnaise made with healthy oil such as Primal Kitchen™ Avocado Oil Mayonnaise or Spectrum Naturals™ Light Canola Mayo.

Dressing ingredients:

1 12-ounce bottle healthy mayonnaise such as the Primal Kitchen™ or
    Spectrum Naturals™ types described above

5 tablespoons fresh or frozen and thawed lemon juice

4½ tablespoons red wine vinegar

2½ teaspoons Dijon mustard

1 tablespoon anchovy paste

¼ teaspoon salt

¼ teaspoon black pepper

⅝ cup grated Romano cheese

Additional salad ingredients:

Romaine lettuce, torn into bite-sized pieces

More grated Romano cheese

Canned anchovies cut into small pieces

To make the dressing, combine all of the dressing ingredients in a blender or food processor bowl and blend until smooth, combine them in a bowl or 4-cup measuring cup and blend with a hand blender or mixer, or beat them together thoroughly by hand. This dressing can be made ahead and stored in the refrigerator.

At mealtime, in a large serving bowl or individual bowls, combine the lettuce with enough dressing to coat it well. Stir in the anchovy pieces. Then sprinkle the top of the salad with the additional grated Romano cheese. Makes about 2 cups of dressing and about 15 servings of salad. Contains zero carbohydrate units.

Condiments

The yeast-free (vinegar-free) condiments below will make your salmon or other burgers special. For vinegar-free pickles, purchase lactofermented pickles from the refrigerated section of your health food store. They are made tangy by fermentation by *lactobacilli* so are good for your intestinal flora and acceptable on diets for yeast allergies. If you would like to make your own lactofermented pickles or other vegetables, see *Healing Basics* as described on page 227.

Homemade Ketchup

1 6-ounce can tomato paste
⅓ cup apple juice concentrate, thawed OR ⅓ cup water plus ¼ teaspoon
    Berlin Seeds™ stevia extract or pure monk fruit extract
3 tablespoons vinegar or lemon juice OR 3 tablespoons water plus
    2 teaspoons tart-tasting unbuffered vitamin C powder
¼ teaspoon dry mustard
Pinch of allspice
Scant 1 teaspoon salt
¼ cup water (optional, or you can add more or less water to make the
    ketchup the thickness you prefer

Stir together all of the ingredients in a small saucepan. Heat over low to medium heat until it begins to simmer. Simmer, stirring every 5 minutes to prevent sticking, for 20 minutes or until it begins to thicken. Add the optional water if it is thicker than you prefer. Makes about 1 to $1^1/3$ cups of ketchup or about 21 1-tablespoon servings each of which contains about 1 gram of carbohydrate or about zero carbohydrate units. if made with apple juice. Contains zero carbohydrate units if made with stevia or monk fruit.

No-Yeast Mustard

This tastes like the mustard you are familiar with but does not contain vinegar.

>   2 teaspoons dry mustard
>   1 cup water plus 1 teaspoon tart tasting unbuffered vitamin C powder OR
>   ¾ cup water plus ¼ cup lemon juice
>   3 teaspoons arrowroot, tapioca starch or cornstarch
>   ¼ teaspoon turmeric
>   ½ teaspoon salt (optional)

Combine the dry mustard and water in a saucepan and allow them to stand for 10 minutes. Add the lemon juice (if you are using it), starch, turmeric and salt. Whisk and heat the mixture over medium heat, stirring it often, until it thickens and boils. Allow to cool briefly. Whisk in the vitamin C powder (if you are using it) and refrigerate the mustard. Makes about 1 cup of mustard or 24 2-teaspoon servings each of which contains zero carbohydrate units.

Miscellaneous Recipes

Most of the recipes below are used in other recipes in this book. The stevia, monk fruit and caramel color working solutions are needed for "Sugar-Free Cola" on page 91. The stevia and monk fruit solutions can also be used to sweeten coffee, tea, or other beverages. The carob and cacao chips are used in cookies and GORP.

The final recipe, "Cacao Candy," is extremely high in phytonutrients, magnesium and other eye-healthy nutrients and is also good for mood improvement.

Stevia Working Solution

>   6 teaspoons (2 tablespoons) Berlin Seeds™ stevia extract (See "Sources,"
>   page 216)
>   ½ cup water

Measure the water into a 2-cup measuring cup. Add the stevia and whisk until it is completely dissolved. Place in a small glass jar and refrigerate. Contains zero carbohydrate units.

Monk Fruit Working Solution

>   7 teaspoons pure monk fruit extract (See "Sources," page 217)
>   ½ cup water

Measure the water into a 2-cup measuring cup. Add the monk fruit extract and whisk until it is completely dissolved. Place in a small glass jar and refrigerate. Contains zero carbohydrate units.

Caramel Color Working Solution

⅓ cup hot water
¼ cup caramel color powder (See "Sources" page 213).

Heat the water until very hot, just under the boiling point. Add the caramel color powder and whisk, whisk, whisk until it is all dissolved. Pour into a small bottle and refrigerate. Contains zero carbohydrate units.

Carob Chips or Candy

Milk-free unsweetened carob chips are getting harder to find for purchase[4] and often contain grain sweeteners or other sweeteners. If you can't find any you can tolerate, here is recipe for homemade carob chunks to use in your cookies or GORP or to have as a treat.

1 cup carob powder, sieved
½ cup melted coconut oil or palm oil such as Spectrum Naturals™
     shortening

Generously grease a baking sheet with melted oil or, for easy removal of the chips, line it with parchment paper.

Press the carob powder through a wire mesh strainer with the back of a spoon to break up all lumps. Measure the sieved powder. Melt the oil in your microwave oven or over low heat in a saucepan and measure out the required amount. Stir together the carob powder and melted oil. The mixture should be a very thick paste. Because of the variations in carob powder, the amount needed may vary and you may need to add a little more melted oil or carob powder to achieve the desired consistency.

Spread the carob mixture on the sheet to a thickness of about ¼ inch. Chill in the refrigerator for about 15 minutes and then score[5] the mixture into ¼ inch chips or larger (1 to 1½ inch) pieces for candy. Chill for at least 30 minutes or until it is hard. Remove the baking sheet from the refrigerator. Remove the chips or candy with a spatula, or just pick up the parchment paper and use it to pour the chips or candy into a storage container. Store the chips or candy in the refrigerator.

Makes about 2 cups of carob chips or 16 servings. Each ⅛ cup serving contains 1 carbohydrate unit.

---

4 I recently found an online source of milk-free unsweetened carob chips. See "Sources," page 119.
5 Use a long knife, such as a butcher or chef's knife, to cut the semi-soft carob into strips about ¼ inch wide. Then cut the carob strips in a direction perpendicular to the first cut to form ¼ inch squares.

Cacao or Chocolate Chips

If you want to make enough chips for a batch of cookies, a small batch will be enough. However, I make large batches and keep the chips on hand in the refrigerator so I can make cookies without planning ahead.

Small batch of chips

¼ cup melted cocoa butter

Sweetener of your choice:

> Berlin Seeds™ stevia extract – with cocoa use 2¼ teaspoons or with cacao use 4 teaspoons, or to taste
>
> Pure monk fruit extract – with cocoa use 2½ teaspoons or with cacao use 5 teaspoons, or to taste

⅝ to ¾ cup Dagoba™ or other organic cacao powder or natural cocoa (not Dutch-process or alkalinized)

Large batch of chips

1 cup melted cocoa butter

Sweetener of your choice:

> Berlin Seeds™ white stevia extract powder – with cocoa use 3 tablespoons or with cacao use 5 tablespoons, or to taste
>
> Pure monk fruit extract powder – with cocoa use 3 tablespoons + 1 teaspoon or with cacao use 6 tablespoons, or to taste

2½ to 3 cups cacao powder or natural cocoa (not Dutch-process or alkalinized)

For a small batch of chips, rub a loaf pan, preferably a foil pan, with a solid fat such as butter, coconut oil or palm shortening to grease the bottom of the pan and about ½ up inch the sides. For a large batch of chips, prepare a 9 by 13 inch cake pan, preferably a foil pan.

Melt the cocoa butter in a saucepan on the stove or a microwave oven and measure it after melting. If melting cocoa butter for a large batch of chips in the microwave, use a 2-cup glass measuring cup to melt about ⅞ cup of the cocoa butter pieces until there are only a few small pieces of solid cocoa butter left in the melted cocoa butter. Then add more small cocoa butter pieces to bring the volume up to 1 cup. Stir until the rest of the cocoa butter pieces melt.

Stir the stevia or monk fruit extract powder into the measured melted cocoa butter thoroughly. Thoroughly stir in the cocoa or cacao, about ¼ of the amount at a time, to make a thick paste about the consistency of play dough. Pour or spoon the chocolate mixture into the pan and spread and press it with a spatula or your fingers to about ¼ inch thick. (If you use your fingers, their warmth will assist in forming the chocolate into a thin layer).

Using a sharp knife, score the chocolate into ¼ inch squares with perpendicular cuts to form square chips. Refrigerate it until it is firm, then remove it from a pan and break it into chips. Store the chips in the refrigerator. Contains zero carbohydrate units.

Cacao Candy

If this candy is made strictly for the flavor and nutrients, you can be a lot less fussy than this recipe implies. I've tried pressing the lumps in the cacao against the side of the container rather than sifting and heard no complaints.

> ¾ cup cocoa butter
> 4½-5 teaspoons stevia or monk fruit extract powder, or to taste
> 1¼ cups cacao powder

Prepare plastic chocolate molds, foil pan(s) or flat-bottomed plastic containers by rubbing them with butter or melted coconut oil or palm shortening. If you are not sensitive to chemicals, you can spray them with an oil spray such as PAM™. Silicone chocolate molds do not need to be greased and are easy to remove the candy from.

If you're short on time or refrigerator space or have low vision, rather than using chocolate molds, try using a small silicone ice cube tray with a lid that makes about 38 hexagonal pieces of candy. (See the footnote below[6] for where to get these trays with lids). The trays can be placed in the refrigerator easily and stacked to save space if needed.

Press the cacao powder through a strainer or sift it with a flour sifter to remove the lumps. (A flour sifter makes lump-removal easy). Measure the cacao powder.

Navitas™ cocoa butter is sold as irregular pieces which are convenient for melting. Put about ¾ cup of pieces in a glass measuring cup which holds at least 2 cups in volume. Microwave just until the pieces are almost all melted, stopping the microwave and stirring often. Then add enough pieces of cocoa butter to bring the volume up to ¾ c. Melt the added pieces by stirring. The goal is to have all of the cocoa butter pieces melted but not be too hot. Ideal temperature is when the outside of the cup feels warm but not hot enough to be uncomfortable to hold.

Stir the stevia or monk fruit powder into the melted cocoa butter thoroughly. Stir in the cacao powder about ¼ cup at a time to produce a smooth consistency. Pour the candy into the prepared molds using the spout of the measuring cup. Refrigerate until cold. Unmold the candy and store it in a container in the refrigerator.

Silicone candy molds are easy to unmold by pressing on the outside (shaped silicone side, not the side where chocolate is poured) and pushing the silicone shape

---

6   https://www.amazon.com/Ailzos-Honeycomb-Easy-Release-Spill-Resistant-Containers/dp/B07FPKNM6R/ref=sr_1_3_sspa?s=home-garden&ie=UTF8&qid=1541191232&sr=1-3-spons&keywords=honeycomb+silicone+molds&psc=1

for the chocolate all the way inside out. If you have plastic molds, they work well but must be greased to unmold easily.

To unmold candy from silicone ice cube trays which make hexagonal pieces of chocolate, turn the tray upside down over a plate and bend it up along the diagonal rows. Starting with the outermost row, press firmly on the bottom of each well with a finger to pop the candy out. Then bend it along the next diagonal row and repeat until all the candy is unmolded.

This recipe makes about one cup of molten candy, just what is needed to fill a silicone ice cube tray with a little room in the tray to spare.

The yield of candy is about a half pound. If made in hexagonal ice cube trays, the recipe makes about 38 candies, each containing zero carbohydrate units.

Ultimate Peace

Until now, this book has been about how to improve physical health, especially eye health. However, we are not only bodies; we also have minds and spirits. No part of us is unaffected by the other parts.

When I was young, I received advice for getting along with people that included avoiding conversation about potentially controversial topics such as politics and religion. This seemed like good advice to me unless I knew that the other person was likely to share my views on such subjects.

Then I spent over four and a half years struggling to breathe with asthma that was not relieved by medication. I felt as if my life were threatened on a near-daily basis for much of that time. Three years ago, while still in the midst of uncontrolled asthma, I experienced another threat to my life when I was diagnosed with cancer.

These experiences have made me think about more than just the health of my body. The cancer could limit the time I have to convey any message, so I feel compelled to say important things I want to say *now*. I am about to break the old rule about what subjects to discuss. I am going to write about religion in contrast to *the real thing*.

It seems to me that religion is often an artificial man-made system that touches only the surface of life, sometimes detrimentally. It offers us no real connection, no personal relationship with God. It usually offers a formula, a system of what we must do and not do to achieve heaven. It may make us feel better mentally, but doesn't really cure what ails the soul or spirit. Sometimes adhering to formulas of rules, rituals, special days, etc. makes us think that we are as close to God as is humanly possible. This then serves as an immunization against *the real thing*. I have known people who persistently follow the formula and go through the motions in this kind of situation even though they find them meaningless. Amazingly, a few people in "rules" religions understand *the real thing* and it doesn't bother them that much of what their belief system teaches is at odds with *the real thing*.

Often people who realize that their religion is artificial and meaningless throw away the whole system with both hands. This is what I did at a young age. As a young child, I liked Sunday school where we heard about Jesus. In junior high school, we learned the rules. It was all so artificial that I felt I could not stand it. Finally, after listening to many complaints, my mother said I didn't have to go to the classes or church any more.

In high school I met new friends who were unusual in a way that I liked. For instance, my friend Leslie had such a sweet, calm spirit, she just exuded the peace that I wanted. My friends explained their beliefs to me, gave me things to read and prayed for me. In John 6:44, Jesus says, "No one can come to me unless the Father who sent me draws him." I've heard that the Greek word translated "draw" can also mean "drag." At age 17, I was easy to draw, as my friends were praying and waiting. I watched my father dragged at age 68 in response to my prayers of almost nineteen years. Cancer was the rope around his neck that dragged him. That he finally came to knowledge and acceptance of *the real thing* was a gift of God's mercy to him and to me.

What is the diagnosis of what ails our minds and spirits? There are a myriad of symptoms – guilt, depression, anxiety, etc. – all manifestations of a lack of peace. A peaceful spirit is needed for true physical and mental health.

I once heard that a psychiatrist said, "If I could get rid of my patients' guilt, I soon would have no patients." We all know that we have failed in many ways, ranging from being irritable at our family and friends to more serious offenses against people. We have fallen short of what our consciences tell us we should do.

There is a difference between true guilt and guilty feelings, or false guilt. With true guilt, I really am guilty. For example, my friend does not deserve me lashing out when I am upset about something that she isn't responsible for. I rightly feel guilty because I am guilty. I have wounded my friend and my conscience tells me this should not happen. *The real thing* gives the solution for true guilt.

Religion is what we try to do for God to earn his favor. *The real thing* is what God does for us, with no merit on our part. There is nothing that we have to do except admit our need and inability to do anything for ourselves and accept what He has done for us. When I hurt my friend, it is also an offense against God who tells us to "Love one another." (John 13:34). To achieve inner peace, I need to make amends with my friend and ask for forgiveness. Forgiveness from God has already been bought for us by Jesus; we only have to admit our need for it and accept it.[1]

The real thing brings us into a personal relationship with God. How can this be with our burden of guilt blocking the path? The answer to this question is that God bears the burden of our guilt Himself in the person of Jesus. Since relationships are two-sided, we must respond by confessing our need and asking Him to take the guilt upon Himself and grant us forgiveness and closeness with Him. Each of us has free will to make this decision. Once we admit that we need His forgiveness, he transforms us. "If anyone is in Christ, he is a new creation: old things are passed away; behold, all things have become new." (II Corinthians 5:17).

To learn more about this relationship, receive forgiveness, and experience true peace with God (the most basic kind of peace), I suggest that you do what I did. Go straight to the source. Ask God to reveal Himself if He really exists. You can be totally honest with Him. Tell Him your hang-ups with religion and doubts about Him. (I heard a man say he used many swear words when he did this. It must have been all right because he found *the real thing*. See the books listed below for the experience of others who did this.[2])

Get a Bible, a version that fits you language-wise, and read it with a blank-slate mind to find out whatever you can, even if you disagree.[3] Be willing to be drawn or

---

1 Stott, John R.W. *Basic Christianity*. (Grand Rapids, MI, Willian B. Eerdmans Publishing Company, 1971), 107-136.

2 For the experiences of two people who asked questions, see these books:

    McDowell, Josh. *Evidence that Demands a Verdict, Volumes I and II*. Here's Life Publishing, San Bernardino, CA, 1989.

    Strobel, Lee. *The Case for Christ*. Zondervan, Grand Rapids, MI, 1998.

3 There is a variety of advice given to a person reading the Bible for the first time. Most often the New Testament, usually the Gospel of John, is recommended as the place to start. However, the first

to find out the whole thing is bunk. You may find *the real thing*. Then "the peace of God, which passes all understanding, will keep your hearts and minds." (Philippians 4:7)

Ultimate peace comes from a close personal relationship with God. The ultimate cure is Jesus.

Additional Information

Stott, John R.W. *Basic Christianity*. William B. Eerdmans Publishing Company, Grand Rapids, MI, 1971. Stott was a British theologian who was greatly respected for over 65 years before his death in 2011. This book is short but complete and is for adults of all ages.

Bethke, Jefferson. *Jesus > Religion: Why He Is So Much Better Than Trying Harder, Doing More, and Being Good Enough*. Nelson Books, a division of Harper-Collins Christian Publishing, Nashville, TN, 2013. This book is written by a young man and directed at young adults but may speak to older adults as well.

The Seeker's Bible, New Testament with notes and helps by Greg Laurie. Tyndale House Publishers Inc., Wheaton, IL, 2000.

Stern, David H. *The Complete Jewish Bible*. Messianic Jewish Publisher, Clarksville, MD, 1998, 2016. Available online here with free shipping: https://www.barnesandnoble.com/w/complete-jewish-bible-david-h-stern/1102359994?ean=9781936716845

Internet Resources

Smith, Colin, M.P. "Meet Jesus" sermon series. UnlockingTheBible.org, 2017.

http://unlockingthebible.org/sermon/he-brings-life/
http://unlockingthebible.org/sermon/he-wants-you-to-be-saved/
http://unlockingthebible.org/sermon/he-is-the-savior/
http://unlockingthebible.org/sermon/he-is-god/
http://unlockingthebible.org/sermon/he-can-be-trusted/
http://unlockingthebible.org/sermon/he-gives-strength/
http://unlockingthebible.org/sermon/he-knows-you-completely/

---

chapter of John contains a great deal of symbolic language that can be confusing. Those of Jewish heritage might start with the Old Testament and learn about their Messiah there before moving on to the rest of the Bible. (See where to get *The Complete Jewish Bible* above). When I, at age seventeen, went to a bookcase in my parents' house and removed a twenty-six year old Bible which still had pages stuck together, I read the book of Ecclesiastes first. I recently read about an Iraqi Christian who was drawn by a passage in the book of Amos. Ask God for leading on where *you* should begin.

Appendix A
Insulin Resistance and the Insulin Resistance Diet

Metabolic syndrome, a cluster of symptoms and risk factors for cardiovascular disease and diabetes, was recognized as early as the 1950s. It was defined as the presence of at least three of these signs: abdominal obesity (an apple-shaped body), high blood pressure, high blood sugar, high serum triglycerides, and a low blood level of high density lipoproteins (HDL, or "good" cholesterol). In more recent years, a hormonal condition in which insulin is less effective than normal for getting sugar into cells was found to be the major factor in the development of metabolic syndrome. The name "insulin resistance" was coined for this condition. Currently, as many as three out of four Americans have insulin resistance.[1]

In individuals with insulin resistance, the amount of insulin in the blood is higher than normal. However, this insulin does not function normally for enabling glucose in the blood to enter cells. Because the cells are resistant to the function of insulin, the blood glucose level remains high after meals rather than the glucose being taken up by cells where it is needed. A state of pre-diabetes or moderate diabetes exists. Eventually, the pancreas becomes worn out and unable to continue producing insulin. At this point, the patient has severe diabetes which requires insulin injections.

The outlook for individuals with insulin resistance has improved in the last two decades. In the 1990s, Dr. Cheryle Hart, MD, and Mary K. Grossman, RD, developed a diet to help their patients lose weight. Their **"insulin resistance diet"** lowers and stabilizes blood insulin and blood glucose levels, increases the blood level of HDL "good" cholesterol, lowers high blood pressure and high triglycerides, and **enables patients to lose weight and maintain the loss without the hunger and struggle common with calorie-counting or low-fat diets.**[2]

Anyone who been told "just eat less and exercise more" or has tried to lose weight using conventional weight loss diets may be thinking, "It cannot be possible to lose weight without hunger." **It is possible**. Dr. Hart spends many pages in her book, *The Insulin Resistance Diet*," refuting conventional diet lore with sound science. There are two major reasons this diet works permanently and without causing hunger: (1) High insulin levels produce a feeling of intense hunger. Because the insulin resistance diet keeps insulin levels low and stable, this eliminates hormonally-induced hunger which may lead to uncontrolled snacking and cravings for sweets, refined carbohydrates, etc. (2) Weight gain results when high insulin levels "turn on" or "turn off" the activity of enzymes that control fat metabolism at a cellular level.

---

1 Hart, Cheryle R., MD and Mary Kay Grossman, RD. The Insulin Resistance Diet. (New York, NY: McGraw-Hill, 2008), ix.
2 Hart and Grossman, xi.

The mechanism of insulin's influence on fat metabolism is two-fold. The first is that a high level of insulin activates an enzyme called lipoprotein lipase. This enzyme catalyzes the production of triglycerides from any fatty acids (digested fat units in the form that is absorbed by the intestine) eaten in a meal. Thus, excess insulin promotes storage of fat by our fat cells rather than using fat for fuel after a meal.[3] In a person with normal insulin levels, any recently eaten fats could have been used for energy during the two hours after a meal. **If insulin levels are high, dietary fat is more likely to be stored in the fat cells.**

The second effect of high blood insulin levels is to inhibit the activity of the enzyme triglyceride lipase which breaks down stored fat for use as energy. Thus, **a person who has chronically high insulin cannot mobilize and burn body fat.**[4]

Individuals with higher-than-normal insulin levels have real difficulty losing weight due to the effects of the two enzymes mentioned above. When they find it near-impossible to lose weight, it is not because they lack willpower or have a character flaw. It is because insulin affects two enzymes to cause fat to be deposited and make it difficult for the body to mobilize and burn fat no matter how much exercise is done or how little is eaten. Additionally, the more overweight a person is, the more difficult it is to lose weight because excess adipose tissue causes an increase in the hormone cortisol that closes the "doors" of cells to glucose.[5] Thus, the blood glucose that cannot enter cells and be used for energy is carried to the adipose tissue and stored as fat.

The good news is that the insulin resistance diet is not difficult to follow consistently or long-term. It keeps insulin levels low and stable, which eliminates hunger and enables weight loss, and can even reverse the signs and symptoms of metabolic syndrome. Once normal weight is achieved, it is easily maintained because the weight-controlling hormone leptin becomes functional. Leptin promotes normal weight, even if one overeats, by decreasing appetite and increasing metabolic rate. A overview of the insulin resistance diet is below.

The Insulin Resistance Diet

The goal of the insulin resistance diet is to keep blood sugar and insulin stable and at normal levels. Anyone beginning the diet will likely have unstable levels initially and will experience some hunger which is a sign of high insulin and/or falling blood sugar. When hunger strikes, you should immediately eat a snack that contains protein to return blood sugar and insulin to healthy levels. Starvation and deprivation are the hallmark of high-carbohydrate low-fat and calorie counting diets. An entirely new mindset emerges with the insulin resistance diet which is to correct

---

3 Montingnac, Michel. Scientific Principles: Basic Principle Behind the Montignac Method. http://www.montignac.com/en/scientific-principles/

4 Hart, Cheryle R., MD and Mary Kay Grossman, RD, *The Insulin Resistance Diet,* (New York: McGraw-Hill, 2001, 2007), 5; and www.montignac.com/en/la_methode_scientifique.php

5 Hart and Grossman, 6.

your blood sugar and insulin levels quickly and banish hunger. With no prolonged hunger, there is little need for willpower.

The insulin resistance diet is useful for many other conditions in addition to overweight including diabetes and heart disease. Although the simple glycemic control advocated by ophthalmologists Marc and Michael Rose, MD, is good for those of normal weight, individuals with diabetes, heart disease, or who want to make consistent progress on weight loss may be better off with the formal balancing of carbohydrates and proteins of the insulin resistance diet. Eat protein-containing snacks two to three times per day and follow other guidelines as presented on the next few pages. With either approach, simple or formal, most of the carbohydrates eaten should be low on the glycemic index (GI). You may also include some moderate GI foods in controlled quantities. High GI foods are best avoided except for "a taste" or small amount on rare special occasions. For more about how glycemic index is measured and tables of GI values for foods, see pages 181 to 195.

Glycemic control is the best way to lose weight because (1) rather than feeling continual starvation, hunger is eliminated immediately with protein containing snacks and (2) it enables us to work with the body rather than struggling against the body in the weight loss process, unlike calorie counting diets, which often cause increased fat gain in the long term. To lose weight, we need to normalize levels of the hormones that determine whether the body stores or burns fat. These hormones include insulin, cortisol, and leptin, among others. In optimally healthy people of normal weight, the leptin system raises metabolic rate and reduces appetite if we overeat, thus restoring us to a normal weight. Decreasing your level of inflammation also helps leptin to function normally. Inflammation can be reduced by good blood sugar control, avoiding allergens and consuming anti-inflammatory fats and anti-inflammatory foods. See page 196 to 197 for a list of anti-inflammatory foods.

Hunger is the enemy if one needs to burn fat rather than store it. To lose fat, insulin levels must be kept low and stable by (1) balancing carbohydrates with protein in every meal and snack and (2) eating protein or balanced protein-plus-carbohydrate snacks between meals at least as often as every two to three hours. **If you feel hungry, it means your insulin level is high. The correct response to hunger is to immediately eat a protein snack or a snack of protein balanced with carbohydrate.**

Following the principles below will stabilize blood sugar, insulin, and other hormone levels and enable fat burning and weight loss without hunger. Do not allow the numbers below to make you legalistic about "food math" as if you were on a calorie counting diet. Weigh a portion of the food once when starting this eating plan, and then eat about that amount the next time. See the glycemic index tables on pages 183 to 195 to determine what amount of a specific food is one unit, meaning that it contains 15 grams of carbohydrate (for carbohydrate foods) or 7 grams of protein (for protein foods).

1. **All meals and snacks should contain the correct balance of protein with the carbohydrate in the meal or snack.** Keep carbohydrate intake at or below two units (30 grams, with 15 grams per carbohydrate unit) per meal or

snack or during any two-hour time period. Most of the carbohydrates consumed should have low glycemic index scores. Balance the number of carbohydrate units comsumed (two, each containing 15 grams of carbohydrate, at most) with the same or a greater number of units of protein (7 grams per protein unit). Have a little fat in every meal; include a salad or vegetable dressed with healthy monounsaturated oil especially if the protein food does not provide much fat. Add enough additional protein foods and non-starchy vegetables to satisfy hunger.

2. **Every morning eat a breakfast containing enough protein to satisfy you plus carbohydrates in the correct amount to balance the protein or less.** Additionally, it is important to eat breakfast early, ideally within the first hour after arising.

3. **Eat protein-containing snacks three times a day**, mid-morning, mid-afternoon, and in the evening or at bedtime. They don't have to be large; a handful of nuts will do. Low glycemic index carbohydrates can also be added to snacks in the correct balance with the protein you are eating. If hunger strikes before it is time for the next meal or snack, eat an additional protein snack, but be careful not to consume more than 30 grams of carbohydrate within a two hour period. Eat enough at each snack to satisfy hunger. High-fiber foods such as vegetables and a little healthy fat can help make snacks interesting and satisfying. See pages 160 to 161 for snack ideas.

4. **Make wise food choices to provide needed nutrients.** For more about the foods and nutrients that help macular degeneration, see pages 33 to 39. Also include the anti-inflammatory foods listed on pages 196 to 197 in your diet. Take supplements that provide support for eyes (pages 27 to 28) as well as the nutrients important for control of insulin levels and inflammation such as chromium and omega-3 fatty acids. Eat lots of nutrient rich low-carbohydrate fruits and vegetables, and make sure healthy fats, such as described on page 36 to 38, are included. Fat is a friend with this type of weight loss, so do eat healthy fats in moderation.

5. **Do some moderate exercise** or brisk walking. Intense or prolonged exercise, especially without food, can cause the body retain fat and burn muscle, which will decrease metabolism and make weight loss more difficult. Never exercise when hungry; have a protein-containing snack first. See the "Exercise Right" page here http://www.foodallergyandglutenfreeweightloss.com/exercise_right.html for information about what type and how much exercise might be right for you.

For more information, see *The Insulin Resistance Diet* by Cheryle Hart, MD, and Mary Kay Grossman, RD.[6] For an approach to the insulin resistance diet adapted for those with food allergies or gluten intolerance, see *Food Allergy and Gluten-Free Weight Loss* as described on page 227 or visit the website http://www.foodallerg-yandglutenfreeweightloss.com

Read the next appendix sections to learn about the glycemic index and get started on the insulin resistance diet. Tables of glycemic index values for foods follow on pages 183 to 195.

---

6 Hart, Cheryle R., MD and Mary Kay Grossman, RD. The Insulin Resistance Diet. (New York, NY: McGraw-Hill, 2008).

Appendix B
Using the Glycemic Index

The glycemic index (GI) is a system of scoring carbohydrate foods according to the effect they have on blood sugar levels of volunteers. The glycemic index score of a food quantifies what actually happens to the blood sugar level of real people when they eat that food.

Testing the effect of various carbohydrates on blood sugar levels was first done in the 1980s by Dr. David Jenkins at the University of Toronto in Canada. He tested the effect of a large number of foods on blood sugar levels of many human volunteers, both normal and diabetic. This testing became standardized and led to the development of the glycemic index.

The glycemic index has been clinically proven to be useful in its application to diabetes, weight loss, appetite control and coronary health.[1] It is used in Australia, Canada, the UK, France, Italy, Sweden and other countries. Unfortunately, the United States' medical establishment remains officially opposed to the glycemic index.[2]

Glycemic index testing results correct a common misconception that all starches are "good" complex carbohydrates and sugars are all "bad." For example, baked white russet potatoes and some types of rice have GI scores that are higher than pure glucose and have a more major effect on our blood sugar level. White sugar falls in the intermediate range with a GI score of 68. This is because each sucrose molecule is made of two single sugars, one glucose and one fructose. The fructose must be processed into glucose by the liver, thus slowing the release of glucose from that half of the sucrose molecule into the bloodstream. The GI score of a food cannot be predicted from whether it contains simple or complex carbohydrates or from the scores of foods in the same food category. For example, fruits have a wide range of GI scores; grains have a similar wide range.

Glycemic index scores for foods are determined using the test results from pure glucose as a reference food and comparing the results from glucose to the results from the test food. For standard glycemic index testing, eight to ten volunteers are given a dose of 50 grams of pure glucose. Their blood is drawn and blood sugar levels are measured periodically over the next two hours. For each volunteer, these blood test results are plotted on a graph of blood sugar level versus time, and the area under the curve of the graph is calculated. The test is repeated on two or three occasions and the results are averaged. Then, at another time, the volunteer consumes a portion of the test food which contains 50 grams of carbohydrate. The blood sugar

1 Brand-Miller, Jennie, PhD, Thomas Wolever, MD, Kay Foster-Powell. MND, and Stephen Colaguiri, MD., *The New Glucose Revolution,* (New York: Marlowe and Company, 2003), 31. Also http://www.montignac.com/en/la_methode_scientifique.php and www.montignac.com/en/la_methode_regime.equilibre.php
2 Brand-Miller, et al., 30.

levels are again tested over a two-hour period, plotted on a graph, and the area under the curve of the graph is calculated. This area is divided by that volunteer's average result when glucose was tested and the result of the division is multiplied by 100. The number obtained is the approximate glycemic index score (GI score or GI value) for the test food. This number is averaged with the results obtained for the other volunteers to calculate the GI score for the food tested. These GI tests for various foods have been shown to be reproducible in testing done in many countries around the world. The values obtained are reproducibly the same for both healthy volunteers and diabetics; however, diabetics have their blood drawn for a three hour period after the test meal rather than for two hours.

To control spikes and dips in blood sugar and weight-depositing insulin spikes or chronically high insulin levels, it is best to choose most carbohydrates in the diet from those that are low on the glycemic index with a GI score of 55 or less. Foods with an intermediate score of 56 to 69 can be eaten in moderation. For best blood sugar and insulin control, high GI foods with scores of 70 or above should be eaten only rarely and in small quantities.

However, it is possible to enjoy favorite foods that are high-GI made in the usual way more often by preparing them with healthier recipes that result in a moderate or even low glycemic impact. For example, white bread has a low GI score if made using a traditional long-fermented sourdough process. The availability of two eye-healthy non-nutritive sweeteners, pure monk fruit extract and enzyme-treated stevia, makes producing healthy desserts easy. These sweeteners can be heated, unlike some chemically made non-nutritive sweeteners, and are a great way to make sweet treats with a lower glycemic impact.

In addition to the glycemic index score of a food, the quantity eaten also determines the impact the meal or food has on blood sugar and insulin levels. If you eat two cups of cooked pasta, it will have about twice the effect of eating one cup of pasta, or twice the glycemic load. The tables of GI scores for foods in the next appendix give one-carbohydrate unit amounts of foods to help you determine how much of a carbohydrate-containing food to eat and how much protein is needed to balance the carbohydrate.

Appendix C
Glycemic Index Values for Foods

This table contains glycemic index (GI) values and serving sizes for one-carbo-hydrate or one-protein units of most commonly eaten unprocessed whole foods. Unfortunately, GI scores for many processed foods are not in this table because they vary from brand to brand and at different times as manufacturers change ingredients or change how the food is processed. In addition, there is a lack of GI data on processed foods because GI scores can be determined only by testing using human volunteers, and very little of this is done in the United States.[1] Without testing specific brands or foods made with specific recipes, it is impossible to accurately know a food's GI score because, for example, two brands of whole wheat bread may be made with different ingredients and different leavening systems or rising times. These differences will then affect the glycemic impact of the bread.

In spite of the lack of GI scores for some foods, GI scores can guide our food choices. It is possible to work around the lack of data for bread, crackers, and other foods made from grains by using the GI scores of cooked whole grains in this table plus the information in this paragraph to determine trends which help us make wise decisions about what to eat. For example, notice that whole wheat and white bread share very similar GI scores. This is because most flour is very finely milled with metal rollers; thus whole wheat flour is as fine and rapidly digested as white flour. However, the GI score for bread made from stone ground whole wheat flour is lower. Therefore, you might read labels and purchase or bake bread and crackers containing whole grain stone ground flour of a variety of kinds, such as Bob's Red Mill™ stone ground quinoa flour. (See the quinoa cracker recipe made with this flour on page 128). You also might choose baked goods made with grains which have lower GI scores in the GI scores section for cooked whole grains on page 186. In addition, note the one-carbohydrate unit serving sizes as you decide which grains you wish to eat most often, either as cooked grains or in bread products. When you choose a grain with a larger one-unit serving size, a one carbohydrate unit portion will provide a larger amount of the grain containing food.

In the absence of GI testing on many grain-based foods, the advice given by Dr. Jennie Brand-Miller for choosing a commercially made bread is to use bread that contains a high amount of legume flours (garbanzo or chickpea flour, fava bean flour, soy flour, etc.) and/or that includes high-fiber additions such a psyllium husks, whole grain kernels, nuts or seeds. In the book *AntiCancer,* Dr. David Servan-Schreiber gives similar advice for the type of bread cancer patients should eat. He recommends eating bread with three or more added whole grains such as rye,

---

1 GI scores for Australian brands of bread, etc. are found in the *New Glucose Revolution* series of books. In addition, GI scores from many countries around the world are found on www.mendosa. com/gilists.htm and www.glycemicindex.com .

oat, flaxseeds, etc. or eating sourdough bread, which has a lower GI score than other types of yeast bread due to the acid produced by the *Lactobacillus* fermentation.[2]

Do not permit this table to induce you to weigh your food routinely. If you are uncertain about how large a serving size should be, weigh it once, or only occasionally (especially for rarely eaten foods), and then judge the portion by appearance the next time you eat that food. The best practice is to eat foods that are as close to nature as possible and only use these tables as a guideline for choosing low GI carbohydrate foods and for linking and balancing your carbohydrate and protein foods. Practice listening to your body rather than becoming legalistic about "food math."

The information in this table was derived from several sources.[3] Some values are missing from the data for some foods because either it is not available (i.e. many vegetables have such a low carbohydrate content that they cannot be tested for a GI score) or it does not apply to that food (i.e. GI scores do not apply to foods which contain no carbohydrates such as meat).

I realize the size of the type in this table is quite small but it was not possible to get all the information for each food on a single line using large type. I hope readers with low vision have a family member or friend who does reading for them, or if you can use a computer, there is another option. Contact me using the information at the bottom of page 4 and I will email you a PDF file of this book (no charge). It can be opened with Adobe Reader and, at the top of the screen, the magnification can to be set to the strength needed.

---

2 Servan-Schreiber, David, MD, PhD. *AntiCancer: A New Way of Life*. (New York: Penguin Group, Inc., 2009), 69.

3 Hart, Cheryle R., MD and Mary Kay Grossman, RD, *The Insulin Resistance Diet*, (New York: McGraw-Hill, 2001, 2007), 44-61; Brand-Miller, Jennie, PhD, Kate Marsh, and Phillipa Sandall, *The New Glucose Revolution: Low GI and Gluten-Free Eating Made Easy*, (Cambridge, MA: Da Capo Press, 2008), 224-243; Brand-Miller, Jennie, PhD and Kay Foster-Powell. MND, *The New Glucose Revolution Shoppers Guide to GI Values*, (Cambridge, MA: Da Capo Press, 2009 and 2010); www.montignac.com/en/ig_tableau.php ; www.mendosa.com/gilists.htm ; and *Food Processor for Windows, Version 7.7*, by ESHA Research, Inc., P.O. Box 13028, Salem, OR 97309.

Food	Serving Size	Carb Units	Glycemic Index Score	GI Range

Carbohydrate Foods
Fruits

Food	Serving Size	Carb Units	Glycemic Index Score	GI Range
Apple, raw	1 medium (5 oz.)	1 or 0[8]	36	LOW
Applesauce, unsweetened	½ cup	1	42	LOW
Apples, dried	4 rings (1 oz.)	1	29	X-LOW[9]
Apricots, raw	2 large or 4 small (6 oz.)	1	38[10]	LOW
Apricots, dried	7 halves (1 oz.)	1	30	X-LOW
Banana	½ small (4") or ½ cup slices (2¾ oz.)	1	52	LOW
Blueberries	¾ cup (4 oz.)	1	53	LOW
Cantaloupe	⅓ of a 5" melon (6 oz.)	1	65	MED
Cherries, dark	⅝ cup (3½ oz.)	1	63	MED
Cherries, sour	⅝ cup (3½ oz.)	1	22	X-LOW
Dates, pitted	3 medium (1 oz.)	1	45	LOW
Grapefruit	1 small (3½") or ½ large (6 oz. without peel)	1 or 0[8]	25	X-LOW
Grapes	1 cup (3½ oz.)	1	53	LOW
Kiwi fruit	2 small (4 oz.)	1	53	LOW
Mango	½ medium (3½ oz.)	1	51	LOW
Nectarine	1 medium (5 oz.)	1	43	LOW
Orange	1 3-inch diameter (5 oz.)	1	42	LOW
Papaya	½ large (7 oz.)	1	59	MED
Peach, raw	2 small or 1 large (8 oz.)	1 or 0[8]	42	LOW
Peach, canned in juice	2 small halves (5 oz.)	1	45	LOW
Peach, dried	2 halves (1 oz.)	1	35	LOW

---

[8] Raw apple, pear, peach, plum, and grapefruit don't have to be balanced with protein in a meal or a snack because they are very high in fiber and the carbohydrate they contain is mostly fructose. Therefore, the impact these fruits have on blood sugar and insulin levels is low and slow. However, if these fruits are cooked, juiced, canned, or otherwise processed, the carbohydrate they contain is absorbed more quickly, so they must be balanced with protein. In that case, the serving size given in this table will count as one carbohydrate unit.

[9] X-LOW indicates that the glycemic index score of this food is extremely low (less than 35). These foods are usually included in the low range. On this chart, LOW means a score of 35 to 55, MED means 56 to 69, and HIGH means 70 or greater.

[10] The GI score of raw apricots on mendosa.com is 34. *The New Glucose Revolution: Low GI and Gluten-Free Eating Made Easy* gives it a score of 38 and *The New Glucose Revolution Shopper's Guide* gives it a score of 57 (medium range).

Food	Serving Size	Carb Units	Glycemic Index Score	GI Range
Pear, raw	1 small (4¼ oz.)	1 or 0[8]	38	LOW
Pear, canned in juice	2 halves (5 oz.)	1	44	LOW
Pear, dried	1½ halves (5 oz.)	1	43	LOW
Pineapple chunks, fresh	¾ cup (4½ oz.)	1	66	MED
Pineapple chunks, canned with juice	⅜ cup (3½ oz.)	1	59	MED
Plum, raw	2 medium (4½ oz.)	1 or 0[8]	39	LOW
Plum, dried (prune)	3 medium (1 oz.)	1	29	LOW
Raisins	2 tbsp. (¾ oz.)	1	64	MED
Strawberries, fresh	15 large, 1½ cups (6 oz.)	1	40	LOW
Strawberries, frozen unsweetened	¾ cup thawed (6 oz.)	1	40	LOW
Watermelon, cubed	1½ cups (8 oz.)	1	76	HIGH

Cooked Whole Grains

Food	Serving Size	Carb Units	Glycemic Index Score	GI Range
Barley, pearled	⅓ cup (2½ oz.)	1	25	X-LOW
Buckwheat	½ cup (3 oz.)	1	54	LOW
Millet	⅓ cup (2½ oz.)	1	71	HIGH
Sorghum (milo)	½ cup (3 oz.)	1	39	LOW
Polenta (cornmeal)	⅓ cup (2¾ oz.)	1	68	MED
Quinoa	½ cup (3½ oz.)	1	53	LOW
Rice, brown (most)	⅓ cup (2¾ oz.)	1	50 to 69	MED-HIGH
Rice, brown basmati	⅓ cup (2¾ oz.)	1	58	MED
Rice, white (most)	⅓ cup (2¾ oz.)	1	76 to 98	HIGH
Rice, white, Uncle Ben's™ converted	⅓ cup (2¾ oz.)	1	45	LOW
Rice, wild	½ cup (3 oz.)	1	57	MED
Wheat, cracked (bulgur)	½ cup (3½ oz.)	1	46	LOW
Whole wheat kernels	½ cup (4 oz.)	1	41	LOW

Food	Serving Size	Carb Units	Glycemic Index Score	GI Range

Grain-based Baked Goods
(Bread, Tortillas, Crackers, etc.)

This section probably lacks the information you most want – the one-carbohydrate unit serving size and GI score for your favorite "normal," allergy or gluten-free bread.[11] Lacking that, use the GI scores of cooked whole grains opposite plus the information below to determine trends that you can use in your food choices. Notice that whole wheat and white bread share very similar GI scores because most flour is very finely milled with metal rollers; thus whole wheat flour is as fine and rapidly digested as white flour. However, the GI score for bread made from stone ground whole wheat flour is lower. Therefore, you might read labels and purchase or bake bread and crackers containing stone ground flour of other kinds, such as Bob's Red Mill™ stone ground quinoa flour. Also, notice that the GI scores for sourdough bread are lower, making sourdough bread made with any grain you tolerate a good choice. In addition, note the one-carbohydrate unit serving sizes as you decide which grains you wish to eat most often in bread products. In the absence of GI testing on many grain-based foods, the advice given by Dr. Jennie Brand-Miller for choosing a commercially made low GI bread is to use bread that contains a high amount of legume flours (garbanzo or chickpea flour, fava bean flour, soy flour, etc.) and/or that includes additions such a psyllium husks, whole grain kernels, nuts or seeds.[12]

Food	Serving Size	Carb Units	Glycemic Index Score	GI Range
Bagel, white wheat	½ of a 4" or ⅓ of a 5"	1	72	HIGH
Barley bread[13]	1 slice (1-1½ oz.)	1	43 to 67	LOW-MED
Buckwheat bread[13]	1 slice (1-1½ oz.)	1	47 to 67	LOW-MED
Chickpea bread[13]	1 slice (1-1½ oz.)	1	55 to 67	MED
Corn tortilla	1 thin 6-inch tortilla	1	46 to 52	LOW
Rice bread[13]	1 slice (1 oz.)	1	72	HIGH

[11] Since very little GI testing is done in the United States, American brands are not found in the information that was available for compiling this table. This list of GI scores for breads contains only general information, not information from testing American brands, and scores can vary from brand to brand. The rice bread listed here is of unknown composition so its GI score could easily be higher than 58. Read the label on your bread to insure that your serving contains 15 grams of net carbohydrate (total grams of carbohydrate minus grams of fiber) and thus is the amount of bread to equal one carbohydrate unit. The list of GI scores for cooked whole grains is also woefully incomplete because some very useful gluten-free foods such as amaranth have never been tested.

[12] For more about gluten-free bread see question 19 on the FAQs page from the glycemicindex.com website. http://www.glycemicindex.com/faqsList.php

[13] The GI scores from several alternative grain breads in this section of the table are from these breads made in other countries. The scores came from mendosa.com/gilists.htm and *The New Glucose Revolution: Low GI and Gluten-Free Eating Made Easy.*

Food	Serving Size	Carb Units	Glycemic Index Score	GI Range
Rice cake, puffed	¾ ounce	1	82	HIGH
Rice crackers	¾ ounce	1	91	HIGH
Rye bread, sourdough	1 slice (1-1½ oz.)	1	42	LOW
Rye bread, whole grain[13, 14]	½ to 1 slice (1-1½ oz.)	1	40 to 72	LOW-HIGH
Rye crispbread	½ ounce	1	63	MED
Sourdough wheat bread[13]	1 slice (1-1½ oz.)	1	54	LOW
Spelt bread[13]	1 slice (1-1½ oz.)	1	54 to 74	LOW -HIGH
White bread (average)	1 slice (1-1½ oz.)	1	75	HIGH
Whole wheat bread (average)	1 slice (1-1½ oz.)	1	74	HIGH
Whole wheat bread, stone ground	1 slice (1-1½ oz.)	1	59 to 66	MED
Whole wheat crackers with sesame seeds	½ ounce	1	54	LOW
Whole wheat tortilla[13]	½ 10-inch tortilla	1	30	LOW

Pasta

Pasta may be a low to medium GI food if it is prepared properly; the scores in this table reflect proper preparation. Pasta should be *al dente,* with resistance to the tooth. If it is over-cooked, it will have a higher GI score and more adverse impact on blood sugar levels.

Food	Serving Size	Carb Units	Glycemic Index Score	GI Range
Corn pasta	½ cup (2 oz.)	1	78	HIGH
Mung bean noodles	⅓ cup (2 oz.)	1	59	MED
Rice pasta	⅓ cup (2 oz.)	1	61	MED
Semolina (durum wheat) pasta	½ cup (2 oz.)	1	35 to 59	LOW-MED
Soba (buckwheat) pasta	⅝ cup (2½ oz.)	1	59	MED

Beverages

Food	Serving Size	Carb Units	Glycemic Index Score	GI Range
Apple juice	½ cup (4 oz.)	1	44	LOW
Beer	12 ounces (1 can)	1	66 to 110	MED-HIGH
Carrot juice, fresh	1 cup (8 oz.)	1	43	LOW
Coffee, unsweetened[15]	8 ounces	0[15]	-	-
Cranberry-apple juice, unsweetened	½ cup (4 oz.)	1	52	LOW

---

[14] The GI scores on mendosa.com/gilists.htm were as low as 40 for several types of rye bread.

[15] Diet soda and coffee or tea prepared without sugar or nutritive sweeteners contain no carbohydrate. However, caffeinated beverages and diet sodas should be consumed with food to moderate caffeine's impact on blood sugar stability.

Food	Serving Size	Carb Units	Glycemic Index Score	GI Range
Grape juice, 100% juice	⅜ cup (3 oz.)	1	53	LOW
Grapefruit juice, unsweetened	⅞ cup (7 oz.)	1	45	LOW
Pineapple juice, unsweetened	½ cup (4 oz.)	1	46	LOW
Prune juice	⅓ cup (2⅓ oz.)	1	43	LOW
Rice milk, unsweetened[16]	½ cup (4 oz.)	1	86	HIGH
Soy milk, unsweetened[16]	1 cup (8 ounces)	0[17]	43	LOW
Soda, diet	12 ounces (1 can)	0[15]	-	-
Soda, not diet	4 ounces	1	48 to 58	LOW-MED
Tea. unsweetened[15]	8 ounces	0[15]	-	-
Tomato juice, sugar-free	1½ cups (12 oz.)	1	38	LOW
Wine, dessert type	4 ounces	1	not available	MED-HIGH
Wine, no added sugar	8 ounces	0[18]	-	-

Vegetables

Most vegetables are low in carbohydrate, high in vitamins, phytochemicals and other nutrients, and can be eaten in any quantity you'd like. Therefore, the serving size listed in this section of the table for most vegetables is "to satiety." Many vegetables are so low in carbohydrates that they cannot be tested for a GI score, so the score and GI range data for these vegetables is blank. Vegetables such as potatoes and corn are sources of concentrated carbohydrates, so the portion size must be kept to an amount supplying 30 grams of carbohydrate (2 units) or less and must be balanced with protein. Dried legumes are high in protein and count as protein foods when balancing meals or snacks. The carbohydrate they contain is mostly indigestible so is it all right eat these in a quantity sufficient to satisfy your hunger. If you want a vegetable you do not see here, search for its GI score at mendosa.com/gilists.htm, or if it is not starchy, eat as much of it as is needed to satisfy hunger.

---

[16] Check the labels on rice and soy milk for sweeteners. If they are sweetened, use the net carbohydrate content to calculate the correct serving size, which will be smaller than for unsweetened milk. See the dairy product section of this table for more about the protein content of soy milk. Rice milk contains a negligible amount of protein and a large amount of high GI carbohydrate.

[17] The carbohydrate in unsweetened soy milk comes from soybeans and, like all carbohydrates from beans, is mostly fiber so doesn't need to be counted or balanced with protein.

[18] The carbohydrates in the grapes used to make wine are completely changed to alcohol during fermentation. This alcohol may or may not (depending on the expert) affect blood sugar balance. Although the zero number of carbohydrate units implies that you do not need to balance wine with protein foods, it is advisable to drink wine with a protein-containing snack.

Food	Serving Size	Units	Glycemic Index Score	GI Range
Artichokes	To satiety	-	-	-
Arugula	To satiety	-	-	-
Asparagus	To satiety	-	-	-
Beans, baked	⅓ cup (3 oz.)	1 prot	47 to 55	LOW- MED
Beans, green, wax, etc.	To satiety	-	-	-
Beans, green baby lima	To satiety	-	32	LOW
Beets	To satiety	-	52	LOW
Broccoli	To satiety	-	-	-
Cauliflower	To satiety	-	-	-
Cabbage, any kind	To satiety	-	-	-
Carrots, raw	To satiety	-	16	X- LOW
Carrots, boiled	To satiety	-	41	LOW
Cassava (yucca), boiled	⅓ cup (1½ oz.)	1 carb	94	HIGH
Corn, kernels	½ cup (3 oz.)	1 carb	37 to 46	LOW
Corn on the cob	1 small	1 carb	48	LOW
Cucumber	To satiety	-	-	-
Eggplant	To satiety	-	-	-
Fennel	To satiety	-	-	-
Garlic	To satiety	-	-	-
Green beans, any kind	To satiety	-	-	-
Greenss (collards, kale)	To satiety	-	-	-
Legumes, dried, cooked	⅓ cup	1 prot	22 to 42	X-LOW- LOW
including black, cannellini, garbanzo, kidney, navy, pinto, soy, and white lima beans, lentils and split peas				
Legumes, chana dal	⅓ cup	1 prot	8 to 11	X-LOW
Lettuce, any kind	To satiety	-	-	-
Mushrooms, any kind	To satiety	-	-	-
Onions, any kind	To satiety	-	-	-
Parsnips	To satiety	-	52	LOW
Plantain, boiled	⅓ cup (1¾ oz.)	1 carb	40	LOW
Peas	To satiety	-	45	LOW
Peppers, all kinds	To satiety	-	-	-
Potato, new red, boiled	4 1-inch (5 oz.)	1 carb	59	MED
Potato, russet, baked, boiled	1 small (4½ oz.)	1 carb	76	HIGH
Potato, sweet Jersey, baked	½ medium (2⅔ oz.)	1 carb	94	HIGH
Potato, sweet Jersey, boiled	½ medium (2⅔ oz.)	1 carb	44	LOW
Potato, sweet orange, boiled	½ medium (2⅔ oz.)	1 carb	66	MED
Pumpkin, canned	1 cup	1 carb	64	MED
Spinach	To satiety	-	-	-
Squash, any kind	To satiety	-	-	-
Swiss chard	To satiety	-	-	-
Taro root	¼ cup (2 oz.)	1 carb	56	MED
Tomatoes, raw, cooked, canned, paste, sauce, or puree	To satiety	-	-	-
Turnips	To satiety	-	-	-
Water chestnuts	To satiety	-	-	-
Yam, true (bitter)	½ cup (2½ oz.)	1 carb	74	HIGH

Food	Serving Size	Units	GI Score	GI Range	Grams of Fat/serving

Protein Foods
Dairy and Egg Products

Dairy products make delicious and convenient protein snacks for a healthy eating plan. This table gives approximate grams of fat per serving, but read the label from your cheese, milk, or yogurt to determine how much fat you are ingesting. The fats in dairy products from pasture-fed animals contains a healthy ratio of omega-3 to omega-6 fatty acids so are beneficial to health and do not need to be limited in the diet.

Food	Serving Size	Units	GI Score	GI Range	Grams of Fat/serving
Buttermilk	1 cup	1 prot	-	-	2
Cheese, hard, fat-free	1 ounce	1 prot	-	-	0
Cheese, hard, low-fat	1 ounce	1 prot	-	-	2
Cheese, hard, regular	1 ounce	1 prot	-	-	9
Cheese, mozzarella, part skim	1 ounce	1 prot	-	-	4
Cottage cheese, 1% fat	¼ cup	1 prot	-	-	0.5
Cottage cheese, 2% fat	¼ cup	1 prot	-	-	1
Cottage cheese, creamed	¼ cup	1 prot	-	-	2
Cottage cheese, dry curd	⅓ cup	1 prot	-	-	0
Cottage cheese, nonfat	¼ cup	1 prot	-	-	0
Cream cheese, nonfat	3 tbsp.	1 prot	-	-	0.5
(Turn ahead three pages for values for regular or reduced fat cream cheese).					
Egg substitute	¼ cup	1 prot	-	-	2
Egg, whites	2	1 prot	-	-	0
Egg, whole (large)	1	1 prot	-	-	5
Ice cream, sweetened with corn sweeteners or sugar	½ cup	½ prot/1 carb	51	LOW	4 to 12
Ice cream, no sugar added	1 cup	1 prot	-	-	4 to 12
Milk, cow's, 1% fat	1 cup	1 prot	-	-	3
Milk, cow's, 2 % fat	1 cup	1 prot	-	-	5
Milk, cow's, nonfat	1 cup	1 prot	-	-	0
Milk, cow's, flavored and sweetened	1 cup	1 prot/2 carb	26 to 31	LOW	0 to 10
Milk, cow's, whole	1 cup	1 prot	-	-	8 to 10
Milk, goat's, whole	1 cup	1 prot	-	-	10
Milk shake	⅔ cup	½ prot/2 carb	GI data not available		5
Milk, soy, sweetened	1 cup	1 prot/1 carb	31	LOW	5 to 20
Milk, soy, unsweetened	1 cup	1 prot	-	-	5 to 20
Pudding, sugar free	1 cup	1 prot/1 carb	40	LOW	3 to 8
Pudding, sweetened with corn sweeteners or sugar	½ cup	½ prot/1½ carb	40 to 47	LOW	3 to 8
Ricotta cheese, part skim	¼ cup	1 prot	-	-	5

Food	Serving Size	Units	GI Score	GI Range	Grams of Fat/serving
Ricotta cheese, regular	¼ cup	1 prot	-	-	8
Romano cheese	⅓ cup (1 oz.)	1 prot	-	-	8
Yogurt, flavored, no sugar added	1 cup	1 prot	-	-	0 to 8
Yogurt, plain	1 cup	1 prot	-	-	0 to 8
Yogurt, sweetened with corn sweetener, sugar or honey	1 cup	1 prot/2 carb	14 to 54	LOW	0 to 8

Meat, Poultry and Fish

Since meat, poultry and fish contain no carbohydrate, they may be eaten in a quantity sufficient to satisfy. The amounts given below for "serving size" are usually one ounce or the amount that will yield one protein unit (7 grams of protein) for the purpose of balancing these protein foods with the carbohydrate foods on the previous pages. However, for a meal you will probably eat 3 or 4 ounces of meat; the amount must be equal to or greater than the number of carbohydrate units you eat at that meal, which is two units at the most. If you eat meat from pasture-fed animals, the fat they contain has a healthy omega-3 to omega-6 fatty acid ratio so can be consumed liberally.

Processed meats are usually high in unhealthy fat and may contain sugar or corn sweeteners so should not be eaten on a regular basis. Check the package label to determine one protein unit (7 grams of protein) and how much fat you will consume.

As long as the meat, poultry or fish you are eating is unprocessed and healthily raised, do not be overly concerned about the portion size, but rather listen to your body and eat until your hunger is satisfied. As you can see from the listing for chicken or turkey, removing the skin (as recommended by standard weight loss diets) saves you just one gram of fat per ounce of meat, so don't deprive yourself if you enjoy the skin. Fat in meat slows digestion and makes your meal satisfy you longer.

Food	Serving Size	Units	GI Score	GI Range	Grams of Fat/serving
Bacon, raw weight	1 strip (0.8 oz.)	¼ prot	-	-	3 to 13
Beef, broiled, or ground beef, broiled or pan-cooked and thoroughly drained					
90% lean ground beef	1 oz.	1 prot	-	-	2 to 3
80% lean ground beef	1 oz.	1 prot	-	-	4 to 5
73% lean "regular" gr. beef	1 oz.	1 prot	-	-	7 to 10
Prime rib roast, ¼" trim	1 oz.	1 prot	-	-	8
Round (rump) roast or steak, ¼" trim	1 oz.	1 prot	-	-	3
Sirloin, ¼" trim	1 oz.	1 prot	-	-	3
Well-marbled steak such as T-bone, ¼" trim	1 oz.	1 prot	-	-	6
Bison (buffalo)	1 oz.	1 prot	-	-	1
Bologna	3 pieces (2 oz.)	1 prot	-	-	16

Food	Serving Size	Units	GI Score	GI Range	Grams of Fat/serving
Chicken or turkey, roasted					
Breast meat, no skin	1 oz.	1 prot	-	-	1
Breast meat, with skin	1 oz.	1 prot	-	-	2
Dark meat, no skin	1 oz.	1 prot	-	-	3
Dark meat, with skin	1 oz.	1 prot	-	-	4
Clams (meat only)	1½ oz.	1 prot	-	-	3
Crab or lobster meat	1 oz.	1 prot	-	-	0.2
Duck	1 oz.	1 prot	-	-	3
Elk or venison	1 oz.	1 prot	-	-	1
Ham, baked	1 oz.	1 prot	-	-	1 to 4
Hot dog, all beef	1½ (2¼ oz.)	1 prot	-	-	13 to 16
Hot dog, all turkey	1 (1½ oz.)	1 prot	-	-	8
Lamb leg, ¼" trim	1 oz.	1 prot	-	-	3
Oysters (meat only)	2½ oz.	1 prot	-	-	2
Pork chop, braised, ¼" trim	1 oz.	1 prot	-	-	5 to 7
Salmon, broiled	1 oz.	1 prot	-	-	2
Shrimp (meat only)	1½ oz.	1 prot	-	-	0.5
Trout, baked	1 oz.	1 prot	-	-	1
Tuna, water packed	1 oz.	1 prot	-	-	0.2
Turkey, ground (8% fat)	1 oz.	1 prot	-	-	2
White fish, poached, baked or broiled (cod, flounder, sole, halibut, etc.)	1 oz.	1 prot	-	-	0.2

Nuts, Seeds, and Nut Butters

Nuts, seeds, and nut and seed butters are great carry-along protein snack foods because they require no refrigeration. Almonds have not gone rancid when kept in my car for months. If your goal is to control inflammation, nuts and seeds are good sources of omega-3 fatty acids, so include them in your diet liberally. Although I could not find a GI score for chia seeds, here is information that indicates they can be consumed on a low-GI diet: http://www.mendosa.com/blog/?p=233

Food	Serving Size	Units	GI Score	GI Range	Grams of Fat/serving
Almond butter, natural[19]	3 tbsp.	1 prot	-	-	28
Almonds	1½ oz.	1 prot	-	-	22
Cashew butter, natural[19]	2 tbsp.	1 prot	-	-	16
Cashews	1½ oz.	1 prot	-	-	22
Chestnuts	5 (2 oz.)	1 carb/¼ prot	54	LOW	1
Hemp seeds[20]	4 teaspoons	-	-	LOW	-

---

19 Natural nut butters are made from nuts and salt and contain no sweeteners.

20 An estimated GI range for hemp seeds is here: http://www.formulazone.com/Help.asp?TID=ILUP&IID=075021&food=Hemp%20Seeds

Food	Serving Size	Units	GI Score	GI Range	Grams of Fat/serving
Macadamia nuts	3 oz.	1 prot	-	-	65
Peanut butter, natural[19]	2 tbsp.	1 prot	-	-	16
Peanuts, shelled	1 oz.	1 prot	-	-	14
Pecans	3 oz.	1 prot	-	-	61
Pine nuts	1 oz.	1 prot	-	-	14
Pumpkin seeds	1¼ oz.	1 prot	-	-	7
Sunflower seeds, shelled	¼ cup	1 prot	-	-	16
Tahini (sesame seed butter)[19]	2 tbsp.	1 prot	-	-	16
Walnuts, black	2 oz.	1 prot	-	-	32
Walnuts, English	1½ oz.	1 prot	-	-	28

Other Foods

Fats

The foods listed in this section do not contain significant amounts of protein or carbohydrates so they are not listed in the carbohydrate or protein food sections of this table and the "units" column below is blank. The information here is to help you know the amount of fat you eat. However, don't deprive yourself of healthy fats; avocados, monounsaturated oils and dairy products from pasture-raised animals are great sources of healthy fats and make meals satisfy longer.

Food	Serving Size	Units	GI Score	GI Range	Grams of Fat/serving
Avocado	1 medium (7 oz.)	-	-	-	31
Butter	1 pat (1 tsp.)	-	-	-	4
Cream cheese	2 tbsp.	-	-	-	10
Cream cheese, low fat (Neufchatel)	2 tbsp.	-	-	-	6
Cream – table or light	2 tbsp.	-	-	-	6
Cream – whipping	2 tbsp.	-	-	-	11
Oils, cooking such as	2 tsp.	-	-	-	9
olive, canola, avocado, coconut, palm, safflower, grapeseed, and nut oils					
Salad dressing					
Blue cheese	1 tbsp.	-	-	-	5 to 8
French	1 tbsp.	-	-	-	6 to 10
Italian	1 tbsp.	-	-	-	3 to 13
Other types of dressing – Read the nutrition label to determine the fat content					

Food	Serving Size	Carb Units	GI Score	GI Range	Grams of Fat/serving

Snack Foods

Highly processed snack foods should be avoided by macular degenration patients and should be rare treats eaten in moderate quantities for healthy people. Dr. Cheryle Hart recommends dark chocolate for frequent controlled splurges because of its high level of nutrients that stimulate the production of neurotransmitters. For sugar-free chocolate, use the recipe on page 172. With the information in this section of the table you will be able determine what size serving of a snack food is within the two carbohydrate unit (30 gram) limit for a meal or snack and how much protein you need to balance it.

Food	Serving Size	Carb Units	GI Score	GI Range	Grams of Fat/serving
Chocolate, dark, plain	1 oz.	1 carb	41	LOW	9 to 11
Chocolate, milk, plain	1 oz.	1 carb	41	LOW	8 to 10
Corn chips	1 oz.	1 carb	42 to 74	LOW- HIGH	8
Jelly beans	6 (¾ oz.)	1 carb	78	HIGH	0
Mars™ fun-size bar	1 oz.	1¼ carb	51	LOW	19
Popcorn, popped with oil	2¾ cups (1.1 oz.)	1 carb	65	MED	8
Potato chips	1 oz.	1 carb	51 to 59	LOW-MED	8

Sweeteners and Spreads

Food	Serving Size	Carb Units	GI Score	GI Range	Grams of Fat/serving
Agave[21]	4 tsp. (¾ oz.)	1 carb	19	X-LOW	0
Corn syrup[22]	1 tbsp. (⅔ oz.)	1 carb	115	HIGH	0
Grape jelly	1 tbsp. (½ oz.)	1 carb	52	LOW	0
Honey, single source	3 tsp. (¾ oz.)	1 carb	35	LOW	0
Maple syrup, pure	3 tsp. (¾ oz.)	1 carb	54	LOW	0
Orange marmalade	1 tbsp. (½ oz.)	1 carb	48	LOW	0
Strawberry jam	1 tbsp. (½ oz.)	1 carb	46	LOW	0
Sugars, pure					
Fructose	1 tbsp. (½ oz.)	1 carb	19	X-LOW	0
Glucose	1 tbsp. (½ oz.)	1 carb	100	HIGH	0
Sucrose (table sugar)	4 tsp. (⅔ oz.)	1 carb	60 to 68	MED	0

---

[21] The GI score given for agave is an average of agave scores which vary between different sources. All are in the extremely low GI range, however

[22] This GI score came from www.montignac.com/en/ig_tableau.php .

Appendix D
Anti-Inflammatory Foods

The foods in this section contain a variety of nutrients that moderate inflammation. Nuts, seeds, and high-fat fish such as salmon contain omega-3 fatty acids that dampen inflammation. Yogurt promotes the establishment of friendly intestinal flora and helps normalize immunity. However, most of the foods listed here quiet inflammation because of the wide variety of phytonutrients they contain. Luteolin (found in green bell peppers and possibly other bell peppers) has an anti-inflammatory effect because it blocks the production of interleukin-6, a powerful promoter of inflammation. Green tea is anti-inflammatory due to its high level of catechin polymers, especially epigallocatechin gallate (EGCG).[1] Citrus flavanoids are found in grapefruit and oranges. Darkly colored fruits are potently anti-inflammatory because they contain high levels of anthocyanins. Resveratrol is found in red grapes and red wine. The vitamin A precursor beta-carotene is found in large amounts in carrots, broccoli, and arugula. Celery and celery seed contain over 20 anti-inflammatory compounds including apigenin.[2] In addition to their anti-inflammatory effect, blueberries and green tea contain chatechins which stimulate fat-burning in abdominal fat cells, thus promoting weight loss in the mid-section of the body.[3]

Although there are only four types of nuts and seeds on Dr. Leo Galland's top 40 superfoods list, other types of nuts and seeds also contain healthy fats that dampen inflammation. Individuals with food allergies should consume a wide variety of nuts and seeds rather than limiting themselves to the types listed here to prevent developing allergies to the nuts and seeds on this list.

Include generous portions of the foods below in your diet every day.

Fruits (Best eaten fresh and raw)
    Apples
    Blueberries (Use blueberries frozen without sugar if out of season).
    Cherries (Use cherries frozen without sugar if out of season).
    Grapefruit
    Oranges
    Pomegranates
    Red grapes*

---

1 Galland, Leo, MD, *The Fat Resistance Diet,* (New York: Broadway Books, 2005), 98. Most of the foods on this list come from the "Top 40 Superfoods" list in this book. Those that do not are marked with *.

2 Cook, Michelle. "13 Foods that Fight Pain." http://www.care2.com/greenliving/13-foods-that-fight-pain.html

3 Cook, Michelle. "12 Surprising Reasons to Eat More Blueberries." http://www.care2.com/greenliving/12-surprising-reasons-to-eat-more-blueberries.html

Vegetables
Arugula
Bell peppers
Broccoli
Cabbage
Carrots
Celery*
Leeks
Onions
Romaine lettuce
Scallions
Shitake mushrooms
Spinach
Tomatoes

Nuts and Seeds (Raw, not roasted)
Almonds
Flaxseeds
Sesame seeds
Walnuts

Animal Protein Foods
Egg whites
Flounder
Salmon
Sole
Tilapia
Yogurt, plain (Stir in fruit and a little stevia or monk fruit extract if desired).

Herbs and Spices
Basil
Black pepper
Cardamom
Chives
Cilantro
Cinnamon
Cloves
Garlic
Ginger
Parsley
Turmeric

Beverages
Blueberry juice
Cherry juice
Ginger tea
Green tea
Pomegranate juice
Vegetable juice (mixed or carrot juice)

Appendix E
How To Read Food Labels

Commercially prepared foods can save kitchen time, but we must know whether they contain ingredients that may be harmful to fragile eyes. To determine if a food is safe, turn to the food label for information. Food labels offer two components: the "Nutrition Facts," which are in a box, and the ingredient list. The ingredient list is required by law to list ingredients in order of the quantity used with the ingredient that makes up most of the food listed first.

NUTRITION FACTS

At the top of the nutrition facts box is the serving size and number of calories per serving. Do not obsess about calories. If you need to lose weight, calories are not all-important; control of hormones such as insulin, leptin, etc. is what counts. (For more about this, see pages 177 to 178). Also, the number of calories per serving can be deceptive. To prevent a food from looking too caloric, the manufacturer may reduce the serving size to lower the single-serving calorie count, thus rendering the serving size unrealistic.

The next section of the nutrition facts lists the amount of fat, carbohydrate and protein per serving. For macular degeneration patients, the amount of trans fats should be zero. On food labels zero may

Nutrition Facts	
Imlak'esh Organics, Sacha Inchi	
Serving Size 1/4 cup (1oz / 28g)	

Amount Per Serving	
Calories 170	Calories from Fat 120

	% Daily Values*
Total Fat 13g	20%
Saturated Fat 1.5g	6%
Trans Fat 0g	
Cholesterol 0mg	0%
Sodium 150mg	6%
Total Carbohydrate 5g	2%
Dietary Fiber 5g	18%
Sugars 0g	
Protein 8.5g	

Vitamin A 0%	•	Vitamin C 0%
Calcium 4%	•	Iron 8%

*Percent Daily Values (DV) are based on a 2,000 calorie diet.

not mean zero, however, because government guidelines call two grams of trans fat per day a safe amount to consume. Macular degeneration patients especially should totally eliminate trans fats, but everyone should try to avoid unhealthy fats. Read ingredient lists and eschew any food that lists hydrogenated fats or partially hydrogenated fats. If possible, choose foods made with healthy oils such as olive, avocado and canola or other acceptable oils such as coconut oil.

The carbohydrate section of the nutrition facts lists total carbohydrates, dietary fiber and sugars. If following a glycemic control diet, what matters is "net" carbohydrate (the amount of absorbable carbohydrate). This is calculated by subtracting the grams of dietary fiber from the grams of total carbohydrate. Every 15 grams of net carbohydrate (one carbohydrate unit) should be balanced in the same meal or snack with at least 7 grams of protein (one protein unit). Consuming more than 7 grams of protein with each carbohydrate unit is beneficial for many individuals.

"Sugars" includes both healthy sugars like the lactose in milk and unhealthy sugars such as high fructose corn syrup. Read the ingredient list and look for sugar, corn syrup, corn sweetener, high fructose corn syrup, sucrose, glucose and dextrose. Avoid foods that contain these ingredients.

On the subject of milk, food labels on milk may have an important additional statement about recombinant bovine growth hormone (rBGH) which is now being called recombinant bovine somatotropin (rBST). Cattle are given estrogenic hormones and rBST to increase milk production. Treatment with rBST causes the cows to produce insulin-like growth factor (IGF) which is excreted in their milk and not destroyed by pasteurization. IGF, including our endogenous IGF which is released after eating sugar, stimulates cancer cells.[1] When we consume milk from cows given Monsanto's rBST drug, Posilac™, IGF poses even more of a problem. The FDA approved this drug in 1993.[2] A 1996 International Journal of Health Services report said that milk from cows treated with rBST contained ten times the level of IGF-1 (insulin-like growth factor-1) as milk from untreated cows, and more recent reports say it is as much as twenty times higher.[3] In 1998, the British medical journal *The Lancet* reported that women with small increases in levels of IGF-1 were up to seven times more likely to get breast cancer at a pre-menopausal age.[4] This timeline of events since the introduction of rBST suggests a reason for the increased incidence of breast cancer in young women. Although most macular degeneration patients are older, this information is included here for younger "cooks" and family members.

If milk is organic, the label will say that the milk does not contain rBST. However, some milk that is not certified as organic, including economical grocery store brands, is also free of rBST. The label on my Kroger-brand milk says, "Our farmers pledge not to treat their cows with rBST." However, the FDA requires that these labels also say, "The FDA has determined that there is no significant difference between milk from rBST-treated cows and non-rBST-treated cows." In my opinion, this statement should be ignored. If an economical brand of milk does not say that the farmers pledged not to use rBST, purchase Kroger, Safeway's Lucerne or Target[5] brand milk or read labels at other stores.

Finally, nutrition labels contain a listing of how much protein and a few vitamins and minerals are in one serving. The protein amount is helpful for balancing protein with carbohydrates as described above. The vitamins and minerals are listed as the percentage of the recommended daily allowance (RDA) that they provide for the nutrient. Keep in mind that these percentages are based on recommended daily allowances of nutrients that may be insufficient for individuals with compromised health.

---

1 Servan-Schreiber, David, MD, PhD. *AntiCancer: A New Way of Life.* (New York: Penguin Group, Inc., 2009), 75.

2 O'Brien, Robyn. *The Unhealthy Truth: How Our Food Is Making Us Sick and What We Can Do About It.* (New York, Broadway Books, Random House, 2009), 98.

3 O'Brien, 102.

4 O'Brien, 102.

5 At the time of this writing, Safeway's Lucerne brand milk and Target brand milk also contain a statement about their farmers pledging to not use rBST. Safeway's Value Corner brand milk does not contain the statement and costs $.50 less per gallon than Safeway's Lucerne milk.

INGREDIENT LIST

Commercially made foods are also required to display an ingredient list on their labels with ingredients listed in order of the quantity present in the food. However, not everything added to the food is required to be listed. Additionally, food manufacturers may list ingredients by euphemistic names. This is especially prevalent with monosodium glutamate (MSG), an excitotoxin that harms the retina and brain.[6] (For more about this, see pages 59 to 60). The first section on the list of hidden ingredients, below, is excitotoxins because these are the food ingredients that are most important for macular degeneration patients to avoid. **Read all labels and avoid those that contain monosodium glutamate, aspartame, or innocuous sounding ingredients such as "natural flavor" which may be monosodium glutamate.**

Label-reading is also essential for those on allergy and gluten-free diets because commercially prepared foods may contain hidden allergens or sources of gluten. These are foods or food additives which appear on ingredient lists disguised by unfamiliar names. For instance, if the ingredient list contains maltodextrin, people who are allergic to corn should not eat that food. A list of derivatives of common allergenic foods which can appear as disguised or hidden allergenic ingredients follows the excitotoxins section.

This hidden ingredients list is not exhaustive. Unfortunately, new additives are developed and food names change so there may be new names for monosodium glutamate or ingredients derived from gluten or allergy problem foods. These new ingredients or ingredient names may not be listed on the following pages. If you find an ingredient not listed here, look it up online and learn more about it to try to determine its source.

Manufacturers change ingredients, so we need to re-read labels occasionally. Enjoy whatever safe commercially prepared foods that you find and the time you save preparing them.

Hidden Toxins, Allergens and Food Derivatives to Avoid in Commercially Prepared Foods

EXCITOTOXINS:[6]

Aspartame, the artificial sweetener sold as NutraSweet™

Aspartate, an amino acid not required by the FDA to be listed[6] on ingredient labels

L-cysteine, an amino acid not required by the FDA to be listed[6] on ingredient labels

Monosodium glutamate (MSG), in all its forms which are listed on the next page[6]

---

6 Blaylock, Russell L., MD. *Excitotoxins: The Taste that Kills.* (Santa Fe, NM, Health Press, 1995). Appendix I.

Additives that always contain monosodium glutamate:

Autolyzed yeast
Calcium caseinate
Hydrolyzed oat flour
Hydrolyzed plant protein
Hydrolyzed protein
Hydrolyzed vegetable protein
Monosodium glutamate
Plant protein extract
Sodium caseinate
Textured protein
Textured vegetable protein
Yeast extract

Additives that frequently contain monosodium glutamate:

Bouillon
Broth
Flavoring, unspecified type
Malt extract
Malt flavoring
Natural beef flavoring
Natural chicken flavoring
Natural flavor or flavoring, unspecified type
Seasoning, unspecified type
Stock
Spices, unspecified type

Additives that may contain monosodium glutamate or other excitotoxins:

Carrageenan
Enaymes
Soy protein concentrate
Soy protein isolate
Whey protein concentrate

WHEAT:

Adhesive stamps and envelopes. Do not lick them; apply water with a sponge instead.
Alcoholic beverages made from grains such as beer, whiskey, gin, and some vodka
Bulgur
Candy – some. Wheat flour may be used for dusting during processing.
Cooking oil spray – some
Couscous
Dextrin – some

Flavorings or extracts – Some contain grain-source alcohol.
Flour, durum flour, graham flour, gluten flour, wheat flour, semolina flour
Gluten
Grain-based coffee substitutes such as Postum™
Hydrolyzed vegetable protein (HVP) or hydrolyzed plant protein (HPP) – some
Imitation seafood or sirimi. Some contain wheat starch as a binder.
Medications, prescription or over-the-counter. Some use wheat starch as a
    filler.
Modified food starch or modified starch – some
Monosodium glutamate (MSG)
Pasta, including those made from a mixture of flours such as Jerusalem
    artichoke pasta made with semolina flour added
Poultry, self-basting – some
Processed and canned meats – some
Wheat germ, bran, or berries, cracked wheat
White (grain) vinegar

GLUTEN:

All of the items listed above under "Wheat" plus:
Alcoholic spirits – some. Canadian celiac groups say that distillation prevents
    gluten from entering the final product; American groups are not sure. Ask
    your doctor.
Caramel coloring, imported source – some
Coffee – some. Flavored coffees are the most likely to contain gluten. Freeze-
    dried coffee is usually the safest coffee other than pure unground beans.
    Consult the manufacturer.
Grains in addition to wheat such as rye, barley, spelt, kamut and triticale
Herbal teas – A few contain malt
Malt, malt flavoring, malt vinegar
Oats may or may not be allowed if processed in a gluten-free facility. Ask your doctor.
Rice syrup – some
Soy sauce – some
Vegetable gum, vegetable protein – some
Vinegar – types which are made from grain. Canadian celiac groups say that
    distillation prevents gluten from entering the final product; American
    groups are not sure. Ask your doctor.

MILK:

Asthma inhalers, powder type, contain lactose which may contain traces of milk
Casein, sodium caseinate, or caseinate
Curds
Hydrolyzed vegetable protein (HVP) or hydrolyzed plant protein (HPP) – some
Lactalbumin

Lactoglobulin
Lactose
Medications, prescription or over-the-counter. Some use lactose as a filler.
Milk products such as butter, cheese, cream, etc.
Powdered, evaporated, or condensed milk
Processed and canned meats – some
Whey

EGGS:

Albumin
Egg pasta
Egg products such as powdered or dried egg, egg yolk, or egg white
Egg substitutes such as EggBeaters™
Globulin
Meringue
Ovomucoid
Ovomucin
Ovovitellin or vitellin
Sauces such as mayonnaise, hollandaise, or tartar sauce
Wine – Some wines may be clarified with egg white.

CORN:

Adhesive stamps and envelopes. Do not lick them; apply water with a sponge
    instead.
Alcoholic beverages – some, especially sweet wines
Asthma inhalers and nasal sprays containing the HFA propellant
Baking powder – most contain cornstarch
Caramel coloring – some
Citric acid (made by growing *Aspergillus niger* on hydrolyzed corn starch)
Corn flour, cornmeal, corn oil, corn syrup, corn sweetener
Cornstarch – often used as a filler in supplements and medications
Dextrose
Dextrin – some
Egg substitutes such as EggBeaters™ may contain maltodextrin, etc.
Flavorings such as vanilla may contain corn syrup
Food starch – some
Fructose – most (also called levulose)
Glucose – most (also called dextrose)
Grits
Hominy
Hydrolyzed vegetable protein (HVP) or hydrolyzed plant protein (HPP) – some
Imitation seafood or sirimi. Some contain cornstarch as a binder.
Instant tea – some

Intravenous solution most commonly used contains dextrose (5DW)

Maltodextrin – most

Medications, prescription or over-the-counter. Some use cornstarch as a filler.

Modified food starch – some

Paper and plastic items – Some plastic wraps and plastic or paper cups and
   plates may be coated with corn oil.

Poultry, self-basting – some

Powdered sugar (contains cornstarch)

Salt – Many contain dextrose to prevent caking.

Starch, unspecified source – some

Sugar alcohols such as erythritol, sorbitol, maltitol, etc. are usually made from
   corn. Xylitol usually is made from corn but may be extracted from birch.

Vitamin C – most. Some brands labeled as "synthetic" are actually manufactured
   from corn. Ecological Formulas™ makes tapioca-source vitamin C. See
   "Sources," page 217 for ordering information.

Xanthan gum – Usually produced by growing bacteria on a corn-source base.

SOY:

Cooking oil spray – some

Hydrolyzed vegetable protein (HVP) or hydrolyzed plant protein (HPP) – some

Lecithin

Margarine – Most margarine contains soy oil or lecithin

Miso

Processed meats – some

Shortening – most[7]

Soy flour, soy oil, soy meal, soy milk

Tamari, soy sauce, worcestershire sauce

Tempeh

Textured vegetable protein (TVP)

Tofu

YEAST:

All alcoholic beverages

Asthma inhalers and nasal sprays containing the HFA propellant

Black (fermented) tea[8]

Cheese

Enriched grain products – Most are enriched with vitamins made from yeast.

---

7 For soy-free non-hydrogenated shortening, try Spectrum Naturals™ palm oil shortening. Palm
oil is a naturally saturated but healthy fat, the best source of palmitic acid needed for mitochodrial
function.

8 Other types of tea, including herbal tea, may contain a small amount of yeast or mold as a
contaminant.

Malted products

Soft drinks which may contain fermented products such as root beer and
    ginger ale

Soy sauce and condiments which contain soy sauce

Vinegar (all kinds) and condiments which contain vinegar, such as mustard,
    ketchup, pickles, etc.

Vitamins and vitamin enriched processed foods – some. Hypoallergenic
    vitamins may be yeast-free.

Yeast breads. Sourdough is not yeast-free, but if made with a traditional starter
    contains "wild" yeasts which some yeast sensitive people can tolerate. Ask
    your doctor before trying sourdough that claims[9] to be fermented with wild
    yeast.

Other foods: If your doctor puts you on a yeast-free diet, he or she may also
    advise you to avoid leftovers, fruit juices, mushrooms, dried fruits and
    spices, all types of tea, sugar, and other foods which may aggravate
    candidiasis.

---

9 If you wish to make your own sourdough bread with a "wild yeast" culture (which will contain
wheat flour) see *Easy Breadmaking for Special Diets, 3ʳᵈ Edition* described page 228.

Appendix F
Bread Machines

When considering bread machines, "Which bread machine should I buy?" is a commonly asked question. The answer varies from person to person and from year to year. Like automobile manufacturers, bread machine manufacturers change their products often. Furthermore, the needs of each home baker are different. Ask yourself what you need: Will you be away from home and want to come home to a freshly baked loaf of bread to go with dinner? If so, get a machine with a delayed cycle timer. Food allergies or other dietary requirements are other considerations, as is the amount of bread you or your family will consume. How often do you want to bake? If you have a large family or do not want to bake often, get a larger capacity machine. How much do you expect to use your machine? How much money can you afford to spend? How much money will the machine save you?

For people who must eat special bread such as those who derive blood sugar control benefits from sourdough or are on food allergy or gluten-free diets, the last question is crucial. If by making your own bread, you eliminate the weekly necessity of buying pricey loaves of commercially made special diet bread, paying more for a bread machine may be justified. Since the early 1990s, programmable bread machines have made it possible to make special diet bread with much less effort from the baker. For yeast bread, usually the ingredients are added to the machine, a button or two is pushed, and in a few hours a wonderful treat is ready to eat. Sourdough bread takes about 28 hours but most of this time is not hands-on time for the baker.

Almost any bread machine can be used to make yeast bread with wheat, spelt, kamut or white rye flour or rice flour plus eggs and stabilizers. For yeast bread made using buckwheat, whole rye, quinoa, amaranth, oat or barley flour or rice flour without eggs and stabilizers, a machine on which the baker can control the length of the last rising time and the baking time is needed.

There are two possibilities for controlling the length of the last rising time before the bread bakes. In terms of initial investment in a bread machine, the more economical choice is to buy a machine with a bake-only cycle and a dough cycle that includes rising time as well as mixing time. You can use the dough cycle followed by the bake-only cycle to make your special bread, but you have to be present to stop the dough cycle and start the bake cycle after the bread has risen for just the right amount of time. (This will be close to the time of the last or only rise time in the recipe's day 2 programmed cycle). Machines with bake-only cycles are usually available (although their makers change from year to year) and some low-priced machines can be used to make special breads in this way.

The option for controlling the last rising time that is more expensive (in terms of initial investment, but it will save time and money on bread in the long run) is to buy a truly programmable machine. Some machines are called programmable because they can be programmed for a delayed start to have bread ready at a certain time. However, with a truly programmable machine, the baker can program the length of various parts of the cycle such as kneading time, rising times, and baking time. The correct last rise and baking time are crucial for many allergy and gluten-free breads. A truly programmable machine is required for making sourdough bread completely in the machine. In addition, programmable bread machines can compensate for environmental conditions that affect bread such as altitude.

As with all bread machines, programmable models change every few years. In the heyday of bread machines in the early 1990s, Zojirushi™ was the first to introduce a truly programmable machine with their BBCCS15 model. Welbilt™ made a programmable machine that made round loaves of bread and West Bend™ made the first machine with two kneading bars. However, Zojirushi™ is the only manufacturer that has continued to make programmable machines. In the late 1990s I got a Zojirushi™ Home Bakery model BBCCV20 which I used to make buns plus several loaves of bread every week for over seventeen years including the years when my sons were hungry teenagers. The Zojirushi™ BBCCX20 which I currently own has served well as my primary bread machine for eighteen years and is still going strong.

In about 2010, Zojirushi™ offered a new model, the BBCEC20, which is very similar to the BBCCX20 and previous Zojirushi™ models which have two kneading bars. I purchased this machine about eight years ago and used its three programmable cycles to develop sourdough bread recipes made entirely in the machine using the Fermapan™ wheat and gluten-free freeze-dried sourdough starter. (See pages 139 to 144 for a few of these recipes). Since the BBCEC20 machine is very similar to the BBCCV20 and BBCCX20, I expect it to be a reliable, long-lasting workhorse like the previous models.

The large Zojirushi™ machines currently sold have a horizontally rectangular pan with two kneading bars and make large (2-pound) normal-shaped loaves of bread. The two kneading bars produce a mixing motion that includes both ends of the pan. These machines' standard cycles include basic, whole grain, dough, quick basic, quick whole grain, and quick dough cycles, a cake cycle, a jam cycle, a signal to add raisins for raisin bread on most of the cycles, and "homemade menu" programmable cycles. The machines' memory can store three homemade menu cycles which you can program to the time you want for any part of the cycle, including the last rising time, which is critical for allergy and gluten-free bread baking. Two programmed cycles are used for sourdough: a day 1 cycle that mixes the sponge ingredients and then keeps them warm for 18 to 20 hours, and a day 2 cycle that incorporates the remaining ingredients, kneads and bakes the bread. One of the "rise" times on the BBCEC20 model can go as high as 24 hours. This is ideal for making a sponge for true sourdough. However, the newer BBPAC20 "Virtuoso" machine allows programming of a rise time of only up to 12 hours and incubates the dough at a higher temperature, which is not ideal for sourdough bread. This higher temperature may also cause failure of wheat-free and gluten-free bread recipes.

The BBCEC20 and BBPAC20 models also include a sourdough starter cycle. This cycle is not for true sourdough made with *Lactobacillus* bacteria as well as yeast, but rather enables the fermentation of flour and yeast for two hours and then uses this sponge as a starter for a loaf of bread. This "sourdough" does not provide the advantages for blood sugar control offered by bread acidified by a long *Lactobacillus* fermentation.

The BBPAC20 "Virtuoso" machine has a few unique new features which, in my opinion, are not improvements. One is the machine's gluten-free cycle, which works well with the rice-potato starch recipes included in the manual, but it is a set cycle. Although the breads developed by Zojirushi™ for this cycle are very tasty, they are made with 60% potato starch and 40% brown rice flour. Potato starch is a highly refined starch from a high glycemic index (GI) food; rice is also high GI. In my opinion, a diet containing breads made from these recipes would increase weight gain, a common consequence of gluten-free diets containing many rice-based foods. In contrast to the gluten-free cycle, the homemade menu cycles of Zojirushi™ machines offer the baker the flexibility needed to make many different kinds of special diet breads using a wide variety of wheat, non-wheat grain, gluten-free grain or non-grain flours by allowing variation in any and all parts of the cycle. The homemade menu cycles make it possible to use almost any gluten-free or allergy flour successfully (in the older machines with a normal incubation temperature) and to compensate for factors such as altitude, etc. The variety of breads that can be made enables avoidance of another common consequence of standard gluten-free diets, namely developing an allergy to rice. (This is the reason for the book *Gluten-Free Without Rice* which is described on page 228).

Another feature of the BBPAC20 is that the machine stops whenever the lid is opened, making it more difficult to make some gluten-free breads. Although it is important to keep the lid closed most of the time, I like to assist the initial mixing with a spatula for stiff gluten-free or allergy breads, such as quinoa or amaranth bread, to insure adequate development of the "structure" of the guar or xanthan gum. Also, in machines that do not stop when the lid is up, it is easier to use a spatula to make sure all the flour is incorporated into a sourdough starter in the first minutes of kneading.

When making wheat sourdough with the BBPAC20 and the Fermapan™ freeze-dried gluten-free starter mentioned above, I noticed that after the initial 18-hour fermentation, the sponge was more bubbly in this machine and had liquefied. The dough in the next step of the process also rose much more rapidly. I took the temperature of the dough in this machine and the BBCEC20 (the two machines were running in tandem) with an instant-read thermometer, and the readings were 78°F in the BBCEC20 and 92°F in the BBPAC20. I suspect that the higher temperature allows the yeast to metabolize more rapidly and overwhelm the *Lactobacillus*, thus resulting in a less acidic, less sour, higher glycemic index bread being produced by the BBPAC20 machine. The final hand-shaped loaf made from dough produced by the BBPAC20 was dense and had a coarse texture although it rose in the same place for the same time as an excellent loaf made from dough which came from the

BBCEC20 machine. Additionally, the loaf from the BBPAC20 machine did not taste as sour as usual.

I had been unable to make good amaranth bread with the BBPAC20 machine. The discovery that it incubated bread dough at 92°F explained the problem. I suspect that the higher temperature promoted faster rising which caused the bubbles of gas produced in the bread by yeast to over-expand and then collapse, thus producing a short, dense loaf which was gummy inside.

The final unique feature of the BBPAC20 is a heating element in the lid. This produces wheat bread that is as beautifully browned on the top as on the bottom and sides. However, for types of sourdough, gluten-free and allergy breads that make short loaves, it produces pale loaf tops.

The quick cycles on Zojirushi™ machines can save you time. I use the quick basic and quick whole grain cycles to make spelt and whole wheat bread in a little over two hours. The quick dough cycle only takes 36 minutes. These machines are very mellow kneaders and produce excellent spelt bread as a result.

If you already own a bread machine that is not programmable and does not have a bake-only cycle but does have a dough cycle which includes rising time, you can still use your machine for the hard part of the job of making yeast bread or sourdough bread which is the initial mixing and kneading. For sourdough bread, see pages 137 to 138 for how to use the dough cycle of a non-programmable bread machine for some parts of the sourdough process.

To make allergy or gluten-free yeast bread with a non-programmable machine, measure out your ingredients into the pan and start the dough cycle. At the end of the rising time, remove the dough from the machine, stir or knead it briefly to deflate it, and put it in an oiled and floured loaf pan. Allow it to rise in a warm place until it is just under doubled in volume. Then bake it at 375°F for 30 minutes to 80 minutes. Very dense loaves, such as rye, take longer to bake than, for example, egg-free rice bread. Your bread is done when it pulls away from the sides of the pans, is well browned and sounds hollow when thumped on the bottom of the loaf.

For recipes and more information on making yeast bread with or without a bread machine, see *Easy Breadmaking for Special Diets* as described on page 228.

References

Abel, Robert, Jr., MD. *The Eye Care Revolution: Prevent and Reverse Common Vision Problems*. Kensington Publishing Corp., New York, NY, 1999, 2014.

Anshel, Jeffrey, OD, and Laura Stevens, M. Sci. *What You Must Know About Age-Related Macular Degeneration*. Square One Publishers, Garden City Park, NY, 2018.

Blaylock, Russell L., MD. *Excitotoxins: The Taste that Kills*. Health Press, Santa Fe, NM, 87504, 1995.

Buettner, Helmut, MD, Editor in Chief. *Mayo Clinic on Vision and Eye Health*. Mayo Clinic, Rochester, MN, 2002.

Carpender, Dana. *The Low-Carb Diabetes Solution Cookbook*. Fair Winds Press, Beverly, MA, 2016.

Conrad, Scott, MD. *The Seven Healers*. Rapha-7-ven, Dallas, TX, 2011.

Crittenden, John. *Blind Faith*. Crittenden, 2014.

Hart, Cheryle R., MD, and Mary Kay Grossman, RD. *The Insulin Resistance Diet*. McGraw-Hill, New York, NY, 2008.

Heier, Jeffrey S., MD. *100 Questions and Answers About Macular Degeneration*. Jones and Bartlett Publishers, Boston, MA, 2011.

Mogk, Lylas G., MD, and Marja Mogk. *Macular Degeneration: The Complete Guide to Saving and Maximizing Your Sight*. Ballantine Publishing Company, New York, NY, 1999, 2003.

Rose, Marc R., MD, and Michael R. Rose, MD. Save *Your Sight! Natural Ways to Prevent and Reverse Macular Degeneration*. Warner Books, New York, NY, 1998.

Samuel, Michael A., MD. *Macular Degeneration: A Complete Guide for Patients and Their Families*. Basic Health Publications, Laguna Beach, CA, 2008.

Thompson, Rob, MD, and Dana Carpender. *The Glycemic Load Diabetes Solution*. McGraw-Hill, New York, NY, 2012.

Thompson, Rob, MD. *The Glycemic Load Diet*. McGraw-Hill, New York, NY, 2006.

Sources of Special Foods, Products, Information and Services

COOKING EQUIPMENT

Bread machine, programmable
(Information about bread machines is on pages 206 to 209).

> King Arthur Flour Baker's Catalogue
> P.O. Box 876
> Norwich, Vermont 05055
> (800) 827-6836
> Online contact: https://www.kingarthurflour.com/contact/
> www.kingarthurflour.com

Zojirushi Home Bakery Supreme, model BBCEC20
https://www.kingarthurflour.com/shop/items/zojirushi-home-bakery-supreme-bread-machine

Bread pan, long and narrow to make 1 carbohydrate unit slices

Norpro 10-inch long bread pan - Available from Amazon.com
https://www.amazon.com/dp/B000SSV61G/ref=twister_B00CC3HOIO?_encoding=UTF8&psc=1

Hamburger bun pan

Available from King Arthur Flour, contact information above
https://www.kingarthurflour.com/shop/items/hamburger-bun-and-mini-pie-pan

Measuring spoons for measuring small amounts of stevia and monk fruit extract

> Natizo
> PO Box 37635
> Philadelphia, PA 19101-0635
> Online contact: http://www.natizo.com/contact-us.html
> http://www.natizo.com

https://www.amazon.com/gp/product/B074SYFQ9Z/ref=od_aui_detailpages00?ie=UTF8&psc=1

Mezzaluna, one handed, double blade
https://www.amazon.com/gp/product/B000QJ3TQC/
ref=oh_aui_search_detailpage?ie=UTF8&psc=1

INGREDIENTS

Baking powder, corn-free

Featherweight Baking Powder
The Hain Celestial Group, Inc.
4600 Sleepytime Drive
Boulder, CO 80301
(800) 434-4246
Online contact: http://www.hainpurefoods.com/about_us/contact_us.php
http://www.hainpurefoods.com
http://www.hainpurefoods.com/products/product.php?prod_id=1842

Berries, high nutrient, frozen or dried

Northwest Wild Foods
PO Box 855
Burlington, WA 98233
360-757-7940
Online contact: https://nwwildfoods.com/contact-us/
https://nwwildfoods.com/
https://nwwildfoods.com/product-category/wild-berries/ (frozen) or
https://nwwildfoods.com/product-category/dried-berries/ (dried)

Bone broth: salmon, halibut and the Bonafide meat and poulty bone broths on next page

Wise Choice Market
#58 18th Street
Rouyn-Noranda, Quebec, J9X 2L5
Canada
514-613-1165
Email: info@wisechoicemarket.com
http://www.wisechoicemarket.com
https://www.wisechoicemarket.com/bone-broth/

Bone broth: beef, turkey, chicken and mixed beef-lamb-bison

Bonafide Provisions
7040 Avenida Encinas, Suite 104-295
Carlsbad, CA 90211
760-683-9146
Online contact: http://bonafideprovisions.com/contact/
http://bonafideprovisions.com/
https://bonafideprovisions.com/collections/bone-broth

Cola special ingredients

Cola flavor extract[1]

OliveNation LLC
50 Terminal Street, Building 2, 7th Floor
Charlestown, MA 02129
781-989-2033
Email: support@olivenation.com
https://www.olivenation.com/
https://www.olivenation.com/cola-flavor-extract.html

Caramel color powder

The Great American Spice Company
10451 Northland Drive NE
Rockford, MI 49341
Phone: 877-6SPICE9 (877-677-4239)
Email: information@americanspice.com
https://www.americanspice.com
https://www.americanspice.com/caramel-color-powder/?AdID=21900cm00cm04816

Madagascar Bourbon Vanilla

The Safeway store near us sells this vanilla for $15 for 4 ounces, less pricey than anywhere else. Also available from King Arthur Flour, contact information on page 211.
https://www.kingarthurflour.com/shop/items/madagascar-bourbon-vanilla-extract-4-oz

---

1 A few days before submitting this book, the Olive Nation website did not have cola extract. Please contact me using the information on page 4 to learn what product is a good replacement.

Flavors and extracts, organic, gluten-, corn- and alcohol-free

Vanilla extract

> Frontier Natural Products Co-op
> P.O. Box 299
> 3021 78th Street
> Norway, IA 52318
> 844-550-6200
> Email: customercare@frontiercoop.com
> www.frontierherb.com

https://www.frontiercoop.com/frontier-organic-vanilla-flavoring2-fl-oz/

Citrus oils for flavoring

Available from King Arthur Flour, contact information on page 211

Set of lemon, lime and orange oils:
https://www.kingarthurflour.com/shop/items/citrus-oil-set-orange-lemon-lime-1-oz

Lemon oil:
https://www.kingarthurflour.com/shop/items/lemon-oil-1-oz

Rye flavor powder (gluten-free)

> Authentic Foods
> 1850 W. 168th Street, Suite B
> Gardena, CA 90247
> (800) 806-4737 or (310) 366-7612
> Email: sales@authenticfoods.com
> http://www.authenticfoods.com

http://www.authenticfoods.com/products/item/37/Rye-Flavor

Flour

Almond flour - blanched, finely ground, suitable for baking and economical

> Honeyville Inc.
> 1040 West 600 North
> Ogden, UT 84404
> (888) 810-3212 or (385) 374-9400
> Email: help@honeyville.com
> https://shop.honeyville.com/

http://shop.honeyville.com/products/flours/blanched-almond-flour.html/

Many flours and baking ingredients such as guar and xanthan gum, vital gluten, thick rolled oats and more

> Bob's Red Mill Natural Foods Inc.
> 13521 SE Pheasant Court
> Milwaukee, OR 97222
> 800-349-2173
> Online contact: https://www.bobsredmill.com/contact/
> https://www.bobsredmill.com/

High quality flour including quinoa, amaranth, buckwheat, sorghum, teff, tapioca starch, arrowroot, baking products such as vital gluten, and many more.

Guar gum: http://www.bobsredmill.com/shop/baking-aids/guar-gum.html
Xanthan gum: http://www.bobsredmill.com/shop/baking-aids/xanthan-gum.html
Vital gluten: https://www.bobsredmill.com/vital-wheat-gluten.html
Oats, extra thick rolled, gluten-free: https://www.bobsredmill.com/shop/cereals/oatmeal/gluten-free-thick-rolled-oats.html

Spelt flour

> Purity Foods, Inc.
> 2871 West Jolly Road
> Okemos, MI 48864
> 517-448-2050
> Online contact: https://natureslegacyforlife.com/contact
> https://natureslegacyforlife.com/

https://natureslegacyforlife.com/product/non-organic-whole-grain-spelt-flour-5lb-bag

> Arrowhead Mills, Inc.
> 110 S Lawton Avenue
> Hereford, TX 79045
> 806-364-0730
> Online contact: http://www.arrowheadmills.com/contact-us/
> http://www.arrowheadmills.com/

http://www.arrowheadmills.com/cpt_products/organic-spelt-flour/

Wheat flour for bread baking
Available from grocery stores or King Arthur Flour, contact information on page 211

Bread flour:
https://www.kingarthurflour.com/products/bread-flour/

Whole wheat flour:
https://www.kingarthurflour.com/products/whole-wheat-flour/

White whole wheat flour:
https://www.kingarthurflour.com/products/white-whole-wheat-flour/

Shortening, Spectrum Naturals™ - organic palm oil only (soy-free)

Spectrum Organic Products, Inc.
5341 Old Redwood Highway, Suite 400
Petaluma, CA 94954
800-343-7833
Online contact: http://www.spectrumorganics.com/contact/
http://www.spectrumorganics.com
http://www.spectrumorganics.com/product/organic-all-vegetable-shortening/

Sourdough Culture, Freeze Dried

Florapan™ French Style Sourdough Starter
Available from King Arthur Flour, contact information on page 211
https://www.kingarthurflour.com/shop/items/french-style-sourdough-starter-5g

Sweeteners, natural, low GI or non-nutritive

Agave

Madhava Natural Sweeteners
14300 East I-25 Frontage Road
Longmont, CO 80540
(303) 823-9000
Online contact: http://madhavasweeteners.com/contact/
http://madhavasweeteners.com
http://madhavasweeteners.com/product/agave-amber/

Stevia, enzyme treated white powder, filler-free

Berlin Seeds
5335 County Highway 77
Millersburg, OH 44654
(330) 893-2091

Berlin Seeds' stevia is the most purely-sweet tasting enzyme-treated stevia and is much less expensive than most enzyme-treated stevia. You may order by phone or request a catalogue and order by mail. This company is Amish so there is no website, but they are kind and efficient with phone orders.

Pure Monk Fruit Extract

Lakanto
3549 N University Ave, Suite 120
Provo, UT 84604
Phone: (800) 513-7936
Online contact: https://www.lakanto.com/pages/contact-page
https://www.lakanto.com/
https://www.lakanto.com/pages/monkfruitextract

Lakanto pure monk fruit extract comes in 50% and 30% strengths. The 30% is used in this book. Be sure to get the PURE extract powder, not a mixture of sweeteners. Use the promotion code WHOLESOMEYUM for a 20% discount.

Thickener - Signature Secrets™ Culinary Thickener

Available from King Arthur Flour, contact information on page 211
http://www.kingarthurflour.com/shop/items/signature-secrets-culinary-thickener-8-oz

Unbuffered vitamin C powder, tapioca source, made by Ecological Formulas™ and used for baking and salads

The Vitamin Shoppe
Customer Care Department
2101 91st Street
North Bergen, NJ 07047
(866) 293-3367
Online contact: https://www.vitaminshoppe.com/u/feedback.jsp
https://www.vitaminshoppe.com/
https://www.vitaminshoppe.com/p/cardiovascular-research-vitamin-c-tapioca-2000-mg-150-g-powder/cv-1038

Yeast, active dry and quick-rise, gluten-, corn- and preservative-free
Red Star™ or SAF™ Red yeast in 1 or 2-pound bags
Available from King Arthur Flour, contact information on page 211

Red Star Active Dry Yeast
http://www.kingarthurflour.com/shop/items/red-star-active-dry-yeast-16-oz

SAF Instant Yeast
http://www.kingarthurflour.com/shop/items/saf-red-instant-yeast-16-oz

SNACKS

Crunchy Chickpeas

> Saffron Road
> American Halal Company
> 1177 Summer Street, Suite 304
> Stamford, CT 06905
> 877-425-2587 or 203-961-1954
> Email: info@saffronroad.com
> https://saffronroad.com/

Sea salt flavor: **https://saffronroad.com/our-products/sea-salt-crunchy-chickpeas/**

Kale Chips, Organic

> Made In Nature
> 1708 13th Street
> Boulder, CO 80302
> 800-906-7426
> Online contact: https://www.madeinnature.com/contact/
> https://www.madeinnature.com

Olive and Sea Salt flavor, similar to the recipe on page 162:
https://www.madeinnature.com/organic-snacks/kale-chips/olive-sea-salt-kale-chips

Nuts and nut butter, milk-free unsweetened carob chips

Sprouted nuts and nut butters (easiliy digestable)

> Radiant Life
> 5277 Aero Drive
> Santa Rosa, CA 95403
> 888-593-9595
> Online contact: https://www.radiantlifecatalog.com/contact
> https://www.radiantlifecatalog.com/

Sprouted nuts:
https://www.radiantlifecatalog.com/product/better-than-roasted-nuts/nuts-seeds-butters

Sprouted seeds:
https://www.radiantlifecatalog.com/product/better-than-roasted-seeds/nuts-seeds-butters

Nut butters:
https://www.radiantlifecatalog.com/product/635/nuts-seeds-butters?gclid=Cj0KCQjw5NnbBRDaARIsAJP-YR_to5CY5OOKO_JtsctdLkuHUHVLf2fyr-bA35mm9TMdHc3Rowi_V6pAaAIzFEALw_wcB

Nuts of all kinds, including macadamia nut pieces for drying, and milk-free unsweetened carob chips

Nuts.com
125 Moen Street
Cranford, New Jersey 07016
800-558-6887
Email: care@nuts.com
https://nuts.com/

Macadamia nut pieces: **https://nuts.com/nuts/macadamianuts/pieces.html**
Carob chips: **https://nuts.com/chocolatessweets/carob/carob-unsweetened.html**

Whole grain crackers

Ak-Mak Whole Wheat Crackers
These crackers are often available in grocery and health food stores.

Ak-Mak Bakeries
89 Academy Avenue
Sanger, CA 93657-2104
559-875-5511
https://akmakbakeries.com/
https://akmakbakeries.com/products/ak-mak-cracker/

Wasa Original Crispbread including sourdough rye crispbread
These crackers are often available in health food stores.

Wasa LLC
885 Sunset Ridge Drive
Northbrook, IL 60062
800-924-WASA (1-800-924-9272)
http://www.wasa-usa.com

Sourdough rye crackers:
http://www.wasa-usa.com/products/crispbread/sourdough/

Gluten-free crackers:
http://www.wasa-usa.com/products/crispbread/gluten-free-sesame-sea-salt/

MISCELLANOUS PRODUCTS

CLEANING SUPPLIES

AFM SafeChoice™ SuperClean™, stains and sealers

American Formulating and Manufacturing
3251 Third Avenue
San Diego, CA 92103
619-239-0321
https://www.afmsafecoat.com/

SafeChoice™ SuperClean™ https://www.afmsafecoat.com/products/cleaners-carpet-care/safechoice-super-clean or https://www.greenbuildingsupply.com/All-Products/AFM-SafeChoice-Super-Clean

SafeChoice™ stains and sealers
http://www.afmsafecoat.com/products/stains-sealers

Microfiber cloths for cleaning, 20% polyamide to pick up dust well

VibraWipe™

Email support@vibrawipe.com
https://vibrawipe.com/

https://vibrawipe.com/collections/microfiber-cloth or https://www.amazon.com/VibraWipe-Microfiber-8-Pieces-ABSORBENT-%20STREAK-FREE/dp/B00CFALFXY

AMISH-MADE ALL HARDWOOD FURNITURE

DutchCrafters™ Amish Furniture
3709 North Lockwood Ridge Road
Sarasota, FL 34234
866-272-6773

https://www.dutchcrafters.com

VISION ASSISTIVE PRODUCTS

Maxi-Aids, Inc.
42 Executive Boulevard
Farmingdale, NY 11735 USA
800-522-6294
Online contact: https://www.maxiaids.com/contactus
https://www.maxiaids.com/blind-and-low-vision-store

Talking cooking thermometer
https://www.maxiaids.com/talking-digital-cooking-thermometer

ONLINE INFORMATION

FOODS HIGH IN LUTEIN AND ZEAXANTHIN:

Vegetables: http://foodinfo.us/SourcesUnabridged.aspx?Nutr_No=338

Foods in general: https://www.macular.org/wp-content/uploads/2016/05/lutein.pdf

GLYCEMIC INDEX TABLES: http://www.mendosa.com/glycemic_index.pdf

MACULAR DEGENRATION INFORMATION:
https://www.webrn-maculardegeneration.com/

NUTRITIONAL HELP

When we searched for a clinic where Mark could take IVs, we providentially found a source of excellent nutritional help of all kinds (IVs, diet and supplements) and for problems beyond Mark's eyes (blood pressure, blood sugar control, etc.). Had we begun by searching for naturopaths in general, we would have discovered that most of the online listings of naturopaths are incomplete to the point of yielding zero or only a few listings in the area near us (a large metropolitan area) and omit naturopaths we know are nearby.

The best website for searching for naturopaths is Healthgrades, which has a seemingly complete list of naturopaths when you search by location here:

Healthgrades Directory of Naturopaths
https://www.healthgrades.com/naturopathy-directory

However, a search begun at their homepage: https://www.healthgrades.com/ by typing "naturopathy" in the "doctors, conditions" box and the zip code in the "for treatment near" box lists only those doctors who have received reviews. This search did not list the doctor Mark saw who had experience with IVs for macular degeneration, and would not have led us to a doctor as well-suited to treat him.

Bottom line: I do not have the answer on how to find the perfect person to help you on this journey. Ask everyone you know and search online in a variety of ways. Naturopaths are trained to take a holistic nutritional approach. They do not administer anti-VEGF drugs so lack the profit motive that ophthalmologists might have. However, if you find a holistic eye doctor, email me using the contact information at the bottom of page 4 and I'll send him or her a copy of this book. Please spread the word (see pages 5 and 6) so others have hope and might have an easier search for help.

Index

Recipes appear in *italics*. Informational sections appear in standard type.

Helpful Books

Beating Macular Degeneration With Nutrition gets to the root cause of macular degeneration, "starvation of the retina,"[1] as well as other influences on eye health such as ultraviolet and high energy blue light, toxins, drugs, blood sugar control, inflammation and exercise. It contains 95 recipes, nearly all of which are gluten-free and food allergy friendly yet will be enjoyed by everyone. Also included is a "Sources" section for special foods, products and sources of help. Stop the progression of macular degeneration and save your sight! (The ***Beating Macular Degeneration e-book*** contains live links to items in "Sources" and online references).

ISBN 978-1-887624-23-7. . . . . $24.95
ISBN for the future large print edition 978-1-887624-24-4 . . . . price TBD

Healing Basics: Prevent Cancer or its Recurrence, Achieve Ideal Weight Without Hunger, Build the Foundation of True Health is, as quoted from the allergy doctor who is the unsung hero of *Beating Macular Degeneration*, "what we all should know" to stay or become healthy. This book explores why Americans' health has deteriorated drastically and takes us back to basics of diet and lifestyle to address the root causes of our health problems. It also presents well documented information to help you decide on medical treatment wisely and explore natural strategies to use along with conventional treatment or alone. It contains 157 recipes, 95% of which can be eaten by those who must avoid gluten or food allergens, including "classics" for health such as bone broth (a wide variety for those allergic to chicken, beef, onions, etc.), fermented vegetables, and fermented milks. (The ***Healing Basics e-book*** contains live links to items in "Sources" and online references).

ISBN 978-1-887624-22-0. . . . . $29.95

Food Allergy and Gluten-Free Weight Loss gives definitive answers to the question, "Why is it so hard to lose weight?" It is because we have missed or ignored the most important pieces in the puzzle of how our bodies determine whether to store or burn fat. Those puzzle pieces are hormones such as insulin, cortisol, leptin, and others. Individuals with food allergies or gluten intolerance face additional weight-loss challenges such as inflammation due to allergies or a diet too high in rice. This book explains how to put your body chemistry to work for you rather than against you, reduce inflammation which inhibits the action of your master weight control hormone, leptin, and flip your fat switch from "store" to "burn." It includes 175 recipes and a flexible healthy eating plan that eliminates hunger, promotes the burning of fat, and reduces inflammation.

ISBN 978-1-887624-19-0. . . . . $29.95

---

1 Abel, Robert, Jr. MD. *The Eye Care Revolution: Prevent and Reverse Common Vision Problems.* (New York, NY: Kensington Publishing Corp., 1999, 2014), 170.

The Ultimate Food Allergy Cookbook and Survival Guide: How to Cook with Ease for Food Allergies and Recover Good Health gives you everything you need to survive food allergies. It contains medical information about diagnosing food allergies and options for treatment. The book includes a rotation diet that is free from common food allergens such as wheat, milk, eggs, corn, soy, yeast, beef, legumes, citrus fruits, potatoes, tomatoes, and more. Instructions are given on how to personalize the standard rotation diet to meet your individual needs and fit your food preferences. It contains 500 recipes that can be used with (or independently of) the diet. Reference sections include food family tables, a listing of grain/alternative grain-containing recipes by grain used, and sources for special foods and products.

ISBN 978-1-887624-08-4. . . . .$29.95

Gluten-Free Without Rice introduces you to gluten-free grains and grain alternatives other than rice such as teff, millet, sorghum, amaranth, quinoa, buckwheat, tapioca, arrowroot, potato starch, and more. It gives you over 75 delicious recipes for muffins, crackers, bread, pancakes, waffles, granola, main and side dishes, cookies, and desserts. (Even ice cream cones!) With this book you can cook easily for a gluten-free diet without relying on rice. If you have gluten intolerance, celiac disease or food allergies, this book will make it easier and more enjoyable to stay on your diet and improve your health.

ISBN 978-1-887624-15-2. . . . .$13.95

Allergy Cooking With Ease, Revised Edition. This classic all-purpose allergy cookbook includes all the old favorite recipes of the first edition plus many new recipes and new foods. It contains over 300 recipes for baked goods, main and side dishes (including comfort foods), soups, vegetables, salads, ethnic dishes, and desserts including lots of cookies. There are "kid" recipes ranging from teething biscuits to no-grain cookies that could pass for Oreos™. Although there are several grain or grain alternatives given among the recipes for each type of baked food (muffins, crackers, etc.), this book has more "fun" recipes and does not cater to the rotation diet to the degree that *The Ultimate Food Allergy Cookbook and Survival Guide* does. If you want to make your allergy diet a little more light-hearted, this is the book for you! It also contains an extensive sources section.

ISBN 978-1-887624-10-7. . . . .$24.95

Allergy and Celiac Diets With Ease: Money and Time Saving Solutions for Food Allergy and Gluten-Free Diets provides solutions to both the economic and time challenges you face. It shows how to shop economically, cook without spending all day in the kitchen, stock your kitchen for efficiency and good health, make the best use of your appliances, have good times with friends and family without breaking the bank, get organized, and be able to do this in limited time. This book contains over 160 money-saving, quick and easy recipes for allergy and gluten-free diets that those on "normal" diets will also enjoy. Over 140 of them are gluten-free. It includes extensive reference sections including "Sources" and "Special Diet Resources" sections to help you find the foods you need.

ISBN 978-1-887624-17-6. . . . . $24.95

Easy Breadmaking for Special Diets, 3rd Edition contains over 200 recipes for gluten-free, allergy, heart healthy, low fat, low sodium, yeast-free and diabetic diets. It includes recipes for breads of all kinds including sourdough, tortillas, bread and tortilla based main dishes and desserts. Use your bread machine, food processor, mixer or electric tortilla maker to make the bread YOU need quickly and easily.

Third edition – ISBN 978-1-887624-20-6. . . . . $24.95

Easy Breadmaking first edition bargain book – ISBN 1-887624-02-3 . . . . $9.95

> With the bargain book we will include an insert of pages from the third edition about current bread machines, preparation of sourdough, and sourdough bread recipes.

The Low Dose Immunotherapy Handbook: Recipes and Lifestyle Tips for Patients on LDA and EPD Treatment gives 80 recipes for patients taking low dose immunotherapy treatment for their food and other allergies. It also includes organizational information to help you get ready for your shots.

ISBN: 978-1-887624-07-7. . . . . $12.95

How to Cope With Food Allergies When You're Short on Time is a booklet of time saving tips and recipes to help you stick to your allergy or gluten-free diet with the least amount of time and effort. **Chose a paper copy** which will be shipped with your paper book(s) **or an e-book** which will be emailed to you as soon as your order is received.

> $5.95 or **FREE** with the order of **two** other paper books on these pages

Economize by chosing E-B00KS

All books here are available as **e-books for $9.95 or less**. For more information about a e-books visit **http://healingbasics.life/livesources.html#SCR** .

There is no shipping charge for e-books.

These e-books are for anyone including, the computer-challenged. They are printable, which is helpful for using the recipes. You will not have problems with downloading. No software other than Adobe Reader is needed. Order e-books easily from the webpage above and get quick delivery of e-books in PDF format attached to an email. You can also order e-books by email or phone. See the e-book prices and more information on the webpage above.

Online Ordering

To order online, vist www.healingbasics.life/books.html or the website of your favorite online bookseller.

Bonus Items

Order **any two paper books** and recieve a **FREE** paper copy or e-book of *How to Cope With Food Allergies When You're Short on Time* or a **free e-book** of your choice. E-books are printable.
Order any two shipping-free e-books and receive a **free *How to Cope* e-book plus an additional free e-book** of your choice.

Mail Orders

To order paper books by mail, remove one of the next two pages, fill out the form and mail it and your check to:

Allergy Adapt, Inc.
1877 Polk Avenue
Louisville, CO 80027

Shipping for Mailed Orders

If the order contains books and e-books, this charge is based only on the value of paper books ordered:

Orders up to $9.99 – Add $4.00
Orders up to $34.99 – Add $7.00
FREE SHIPPING on any paper book order over $35

Questions about these books? About computer challenges with e-books? About your order or shipping? Call **303-666-8253** or email **contact@healingbasics.life**.

Thank you for your order!

A paper books order form is on the opposite side of this page.

To order e-books online
visit http://healingbasics.life/livesources.html#SCR

To order by phone or email
call 303-666-8253 or
email contact@healingbasics.net

Paper Book Order Form

Send books to:

Name: _____

Street address: _____

City, State, ZIP code: _____

Email or phone number (for questions about order)**:** _____

Item	Quantity	Price	Total
Beating Macular Degenration With Nutrition		$24.95	
Healing Basics		$29.95	
The Ultimate Food Allergy Cookbook and Survival Guide		$29.95	
Gluten-Free Without Rice		$13.95	
Allergy Cooking With Ease		$24.95	
Easy Breadmaking for Special Diets – First Edition Bargain Book Third Edition		$9.95 $24.95	
Allergy and Celiac Diets with Ease		$24.95	
List other book(s) from previous pages here:			
Circle one bonus item of your choice: *How to Cope* paper book or e-book E-book of your choice - Print title below.		**FREE with 2 book purchase**	
Order any **TWO** paper books and get ***How to Cope*** or the ***e-Book of your choice*** **FREE!**	Subtotal		
	Shipping-See amount on page 230		
	Colorado residents add 8% sales tax		
	Total		

A second copy of the paper books order form
is on the opposite side of this page.

To order e-books online
visit http://healingbasics.life/livesources.html#SCR

To order by phone or email
call 303-666-8253 or
email contact@healingbasics.net

Paper Book Order Form

Send books to:

Name: _____

Street address: _____

City, State, ZIP code: _____

Email or phone number (for questions about order): _____

Item	Quantity	Price	Total
Beating Macular Degenration With Nutrition		$24.95	
Healing Basics		$29.95	
The Ultimate Food Allergy Cookbook and Survival Guide		$29.95	
Gluten-Free Without Rice		$13.95	
Allergy Cooking With Ease		$24.95	
Easy Breadmaking for Special Diets – First Edition Bargain Book Third Edition		$9.95 $24.95	
Allergy and Celiac Diets with Ease		$24.95	
List other book(s) from previous pages here:			
Circle one bonus item of your choice: *How to Cope* paper book or e-book E-book of your choice - Print title below.		**FREE with 2 book purchase**	
Order any **TWO** paper books and get ***How to Cope* or the *e-Book of your choice* FREE!**	Subtotal	
	Shipping-See amount on page 230	
	Colorado residents add 8% sales tax	
	Total	

www.ingramcontent.com/pod-product-compliance
Lightning Source LLC
Chambersburg PA
CBHW080329270326
41927CB00014B/3141